LUMINOUS PURSUIT

Jellyfish, GFP, and the Unforeseen Path to the Nobel Prize

LUMINOUS PURSUIT

Jellyfish, GFP, and the Unforeseen Path to the Nobel Prize

Osamu Shimomura
Marine Biological Laboratory, USA

Sachi Shimomura
Virginia Commonwealth University, USA

John H Brinegar
Virginia Commonwealth University, USA

World Scientific

NEW JERSEY • LONDON • SINGAPORE • BEIJING • SHANGHAI • HONG KONG • TAIPEI • CHENNAI • TOKYO

Published by

World Scientific Publishing Co. Pte. Ltd.

5 Toh Tuck Link, Singapore 596224

USA office: 27 Warren Street, Suite 401-402, Hackensack, NJ 07601

UK office: 57 Shelton Street, Covent Garden, London WC2H 9HE

Library of Congress Cataloging-in-Publication Data

Names: Shimomura, Osamu, 1928– | Shimomura, Sachi. | Brinegar, John H.

Title: Luminous pursuit : jellyfish, GFP, and the unforeseen path to the Nobel Prize / by Osamu Shimomura (Marine Biological Laboratory, USA), Sachi Shimomura (Virginia Commonwealth University, USA), John H. Brinegar (Virginia Commonwealth University, USA).

Description: Hackensack, NJ : World Scientific, 2017. | Includes bibliographical references and index.

Identifiers: LCCN 2016059282| ISBN 9789813202153 (hardcover : alk. paper) | ISBN 9789813202160 (softcover : alk. paper)

Subjects: LCSH: Shimomura, Osamu, 1928– | Marine biologists--Massachusetts--Biography. | Marine biologists--Japan--Biography. | Nobel Prize winners--Biography.

Classification: LCC QH91.3.S55 .S57 2017 | DDC 578.77092 [B] --dc23

LC record available at https://lccn.loc.gov/2016059282

British Library Cataloguing-in-Publication Data

A catalogue record for this book is available from the British Library.

Typeset by Stallion Press
Email: enquiries@stallionpress.com

Printed in Singapore by Mainland Press Pte Ltd.

Dedication

To friends, family, and all those who helped. I hope they will enjoy my story, which is sometimes theirs too.

Preface

My path to the Nobel Prize was full of unforeseen events and circumstances: scientific study, people, and chances that led me to unimagined results—just like winning the prize itself, which I never expected through my many years of research. So this is a story about serendipity and about the outcomes of studying nature at the right time, especially my quest to understand the science behind the mysterious animals that glow with brilliant light in deep oceans or dark nights. It is also a true story about pursuing pure science without a fixed end in mind, and learning from both experiences and experiments, with the help and encouragement of many generous and memorable people along the way.

When I was a child and my family moved to join my father, an army officer, in Manchuria, I could not foresee how World War II would shape my future and channel my education into the only paths available to me after the war. When, as a high school student working in a Japanese military factory, I saw the atomic bomb fall on Nagasaki, I could not guess when or where I would be able to resume the process of acquiring an education, and certainly never expected to be a scientist in America one day. When I took my mentor's advice to undertake work on sea fireflies (*Cypridina*), I was determined to find a solution to the scientific questions they posed for me, but had no idea then that my studies would lead me to my life-long work in the field of bioluminescence. When I came to the United States as a

Fulbright Scholar in 1961, I never expected to be living here still, five and a half decades later. When I started researching the living light produced by jellyfish, and took my family to scenic San Juan Island to collect buckets upon buckets of the luminescent jellyfish *Aequorea* each summer for over a decade, and helped craft mechanical devices out of household appliances to speed our processing of jellyfish, I did not expect to discover the Green Fluorescent Protein (GFP).

When I purified GFP, I did not expect it to help other scientists make further discoveries, create a rainbow palette of fluorescent proteins, and even produce luminous green pets. By now the effects of my work have spread beyond the laboratory: jellyfish are popular stuffed animals in Japan (I have received several as gifts myself), and my six-year-old granddaughter is proud of how well she can draw jellyfish—"my Japanese grandpapa showed me how." I would not have expected any of this to come to pass from the shifting tides of events. Certainly I never foresaw that results from my study would become so important in medical research, and even bring myself and my colleagues in GFP research, Martin Chalfie and Roger Tsien, the Nobel Prize in Chemistry when I was eighty years old.

Numerous people aided our pursuit of the elusive secrets of bioluminescence, over the years, or helped in other ways. Many are mentioned in this book. I hope all of them understand my gratitude. A Japanese version of my story, *Learning From Jellyfish [Kurage ni manabu]*, was published by Bunkensha (Nagasaki, 2010). With this English rendition, my children and grandchildren can know more about their pasts; this book is for them too.

I thank the Nobel Foundation. This version of my story expands upon events, anecdotes and chronology in my Nobel biography and lecture. For providing memorable vignettes and photos, I thank Thomas P. Carroll, who allowed me to quote his letter to me, and Diana Kenney at Marine Biological Laboratory for her kind help in locating photographs held by MBL. For faith in this project, which we might not otherwise have undertaken, and her work on translating the first chapter, I thank Dr. Reiko Matsui. For many helpful suggestions regarding language, readability, and clarity, my co-authors and I thank Jayne Harding and Cristina Stanciu at Virginia Commonwealth University. Last but not least, I thank relatives and friends for their support and for all their collection of luminous species.

Osamu Shimomura

Contents

Dedication		v
Preface		vii
Prologue		1
Chapter 1	My Early Life: Manchuria, Sasebo, and the Beginning of the War	5
	1.1 From Fukuchiyama to Sasebo	5
	1.2 Shimomura Family Traditions	8
	1.3 In Manchuria	12
	1.4 To Sasebo Again	15
	1.5 In Osaka	20
Chapter 2	The War Strikes Home: Ohmura Naval Aircraft Factory and the Atomic Bomb	23
	2.1 Student Mobilization Order at Isahaya Middle School	24
	2.2 The Atomic Bomb on Nagasaki	27
	2.3 After the War: Searching for a Future	29
	2.4 After the War: Disappointments	32
Chapter 3	Struggling to Learn: Nagasaki, Nagoya, and Luciferin (1948–57)	35
	3.1 Nagasaki Pharmacy College and Nagasaki University	35

	3.2 The Hirata Lab and *Cypridina* Luciferin	41
	3.3 An Unexpected Success: Luciferin Crystals	46
Chapter 4	To America and Back: Marriage, Princeton, and Discovery of Aequorin and GFP	51
	4.1 Marriage with Akemi: Departure from Japan	53
	4.2 To Princeton	59
	4.3 Friday Harbor	65
	4.4 The Jellyfish *Aequorea*	71
	4.5 The Busy Year of 1962: Finding the First Traces of GFP	74
	4.6 Fireworms in Bermuda	77
Chapter 5	A Risky Endeavor, Raw Jellyfish, and Mysteries of Bioluminescence	81
	5.1 Back in Nagoya: 1963–65	81
	5.2 Glow Worms in New Zealand	83
	5.3 Back to Princeton: My Family Grows	87
	5.4 Luminescence Mechanism of Aequorin: Friday Harbor	95
	5.5 How Do You Prepare Jellyfish?	97
	5.6 My Research on Subjects Other Than Aequorin: 1965–1978	103
	5.7 Photoprotein	108
Chapter 6	From Mushrooms in Woods Hole to Red Jellyfish in Bergen	111
	6.1 Dr. Johnson's Retirement: Our Move to MBL (1977–81)	112
	6.2 At the Marine Biological Laboratory in Woods Hole	115
	6.3 Tsutomu and Sachi Leave Home	118
	6.4 My Friend Goto	128

Chapter 7 The 1990s: Losses, Gains, and the Rise of GFP 131
7.1 Dr. Johnson's Passing 131
7.2 The Passing of Relatives in Japan 133
7.3 The Value of Pure Science 139
7.4 Norway and the Jellyfish *Periphylla* 140

Chapter 8 Going Full Circle: Nature's Gift, Scientific Honors, and Revisiting Japan 145
8.1 After My Retirement 145
8.2 Nature's Gift: Abundance and Disappearance of *Aequorea* 148
8.3 More Honors, The Asahi Prize, and Its Influence 152

Chapter 9 The End of the Path: The Nobel Prize Announcement and Nobel Week 157
9.1 Preparing for Stockholm; the Order of Culture Award 158
9.2 To Stockholm 164
9.3 The Nobel Week Begins 166
9.4 The Nobel Prize Ceremony and the Nobel Banquet 170
9.5 The End of Nobel Week 173

Chapter 10 After Stockholm: Visiting Los Alamos, Remembering Nagasaki 177
10.1 My Welcome in Japan 178
10.2 Lindau, Germany 180
10.3 Research in Russia 182
10.4 An Invitation to Los Alamos, Where the Atomic Bomb was Built 185
10.5 Fragments of History? 188
10.6 Akemi's Memories of the Atomic Bomb 193
10.7 Science and War: Back to Nagasaki 198

Chapter 11 Epilogue 203
11.1 Was the Discovery of GFP Good Luck? 204
11.2 My Research Life and Communications with Other Scientists 204
11.3 The People I Respect 206
11.4 A Word of Thanks 207

Selected Bibliography 209

Index 211

Prologue

Nobel Prize phone call from Stockholm, 5:15 a.m. EST, 8 October 2008

The telephone rang. It was 5:15 in the morning. My wife Akemi picked up the handset, and said, "It is from Stockholm." I had no choice but to take the call although my brain was still foggy and nearly asleep. I heard a woman's voice, but the connection was awfully noisy; a strong echo kept me from hearing her words distinctly. After she said something, a man spoke, reading a long sentence. Judging by the words that I could hear, I understood *I am going to get the Nobel Prize in chemistry together with Martin Chalfie and Roger Tsien*. My first reaction at that moment was that I was getting into terrible trouble: If I accepted a Nobel Prize, I would be targeted by the media and my life as an experimental scientist would end. However, if I declined the prize, I would disappoint people in Japan, who would condemn me for refusing such an honor for my country, bringing me more serious trouble. I realized I had no option but to accept it.

By the time a third person came on the phone, the noise had lessened enough so that I could hear him say, "We will make the announcement to the media twenty minutes from now. Then you should expect a rush of phone calls." I had never in my life thought about such a miracle happening—that a person like me, who works in such a minor field of science as bioluminescence, could get a Nobel Prize. It was completely a bolt out of the blue.

I wanted to call Professor Hirata's wife immediately to let her know the news, because the late Dr. Hirata, my mentor, had trained several students into top world-class chemists, and I knew that Mrs. Hirata was hoping for the Nobel Prize for them very much. However, Akemi told me to have breakfast first, to prepare for the hard day ahead. She went down to the kitchen to prepare some toast and tea. I washed my face and sat down to have my breakfast. But, before I had time for a single bite, the phone rang again. Akemi answered. It was a call from the *Asahi Shimbun*, the major Japanese daily newspaper; my conversation with them lasted nearly thirty minutes. More calls soon followed, arriving almost continuously from various media companies including TBS, Fuji TV, TV Asahi, Japan TV, and a German radio station. While Akemi was hurrying to answer phone calls, someone knocked at the door. It was a photographer from Reuters. I let him into my office, he took some pictures quickly, and then he left within a few minutes. Next came the Associated Press photographer.

At 8:30, Pamela Clapp showed up at our house. At that time the Director of Development and External Relations of the Marine Biological Laboratory (MBL) in Woods Hole, Massachusetts, I had known her for many years. She instructed me to arrive at MBL in two hours for the press conference at eleven o'clock. She offered to switch all calls coming to my home to MBL, but after a little thought I had to decline her offer because I expected several of the calls to be in Japanese. When I left my home in my car at 10:15 in the morning, there were reporters overflowing the area in front of my house. The ten-minute drive through the back roads to Woods Hole gave me my only brief period of relative quiet. Upon my arrival at MBL, Naoki Noda, whom I had previously helped in making GFP crystals, was waiting for me; he guided me to a parking spot specially reserved for me, then escorted me to the conference room.

There were about one hundred reporters at the joint press conference. The majority were Japanese reporters, probably because a large number of Japanese journalists had been in Boston, about a ninety-minute drive away, to cover the Red Sox game. I found the press conference a little bewildering, however, because of the mixture of Japanese and English being relayed to Tokyo where some of the conferees were. Although I had lived in the United States since the 1960s, it was still a little tricky to switch languages so rapidly because I had never before prepared for such a situation. (There are stories of ancient generals or statesmen able to hold several simultaneous conversations in different languages, and perhaps my very articulate son could do so more glibly, but I found this situation confusing and exhausting.) In answering reporters' questions, I emphasized that my winning a Nobel Prize was totally unexpected, especially in chemistry—I had thought I would be a more likely possibility for the physiology prize, if my name were ever to come up for consideration. I also stressed the fact that all previous Japanese Nobel Prize winners in science were graduates from top former imperial universities, whereas I was a graduate of a provincial college located at the far end of Japan.

After the press conference, I gave about twenty phone interviews, five minutes each, at Candle House (the stone building that houses the main MBL office), and I was finally released at four in the afternoon. When I got home, the winding, wooded road in front of my house was nearly blocked by three big television broadcasting satellite relay trucks. Later our neighbors said they got into a traffic jam in the usually quiet street. Three parties were still waiting for me at my home to interview me, and NHK (Japanese public television), Japan Television, and Fuji Television were all preparing for television relay broadcasting. Off to one side, I received an urgent request from the Japanese Ministry of Education; it said, "The Prime Minister, Taro Aso, will call you between 7:45 and 8:15 tonight. Please keep your telephone line open for the call." We have only one phone line, and that was continuously being called from outside. So it was a very difficult request to fulfill. But, fortunately, exactly at 8:00 p.m. (Japanese time 9:00 a.m.) the Prime Minister's call was connected. Prime Minister Taro Aso said, "Thank you for giving cheer to the Japanese nation by winning the Nobel Prize"; by his tone, he sounded as if he were reading the words from a page. To their later disappointment, reporters in the same room missed their chance to take pictures of me while I was talking with the Prime Minister.

My son and daughter were finally able to reach me by phone just before 8:00 p.m. and I spoke just a few words with them, telling them of the request from the Japanese government to keep the phone line clear. My daughter later told me that she first realized that I had received the Nobel Prize when her brother (also a scientist) emailed her early that morning, followed by several emails from friends in the United States, Italy, and Japan who had heard the announcements of the award. My own brother and sister in Japan learned about my award from the television or from relatives who saw the news on television—neither Akemi nor I had had time to call anyone or to check our email that day.

The last television interview—mostly live broadcasts on Japanese morning news—concluded after 10:00 p.m. The day was finally over. It was, however, only the first day of the busiest days of my life that followed for many months.

CHAPTER 1

My Early Life: Manchuria, Sasebo, and the Beginning of the War

I was born in 1928, in the Year of the Dragon, at Fukuchiyama, within the Kyoto prefecture, in Japan; my brother Sadamu was born four years later in the nearby town of Ayabe. The two of us were raised in this area until I was five years old. My father's name was Chikara, meaning "strength" in Japanese. He was either an army lieutenant or captain of the Fukuchiyama regiment—I do not remember which rank.

1.1 From Fukuchiyama to Sasebo

I remember only two things distinctly from this period. The first happened one morning when I was watering with an extension hose—I don't recall whether it was on the flowers in the garden or on the dusty dry road in front of my home that led to the regiment area. Then, suddenly the hose jerked and bent, seemingly by itself, drenching a fully uniformed officer who was just passing by on the road. The officer glared at me with angry eyes. I ran back home and told my mother about what had happened. Several years later, I learned that my father had apologized to that officer, taking with him a bottle of *sake*.

The other thing I still remember occurred on a Sunday. My father took me to Amanohashidate, a long wooded sandbar formed in a bay, which is famous for its beautiful scenery. It is considered to be one of the traditional "three views" of Japan. However, I did not enjoy that outing at all. The train to the place was very crowded, so that we had nowhere to sit. Moreover, when we got there, I did not find the famed view particularly inspiring. Amanohashidate means "bridge in heaven," and the scenery is supposed to be at its best if you turn your back, bend over, and view it between your legs. This is said to make it look as if the sandbar bridge is floating in the heavens, and may also bring a viewer good luck. I tried it, but the view didn't look any better to me than it had right side up. As for good luck, it seems it took a long time to arrive.

Soon after birth, with my grandmother Tsuki, mother Yukie, and the midwife (1928).

Many years later when I had children, I found that they do not enjoy glorious scenery as much as adults do. When my son Tsutomu was young, he was not interested in the landscape at all when we stopped at scenic locations while traveling; instead of looking at the view, he used to draw something on the ground with a stick or pebble or peer into the drainpipes by the roadside.

In the spring of 1933, my father took his new post at the independent defense regiment of Renzankan (present Lienshankuan) in Manchuria, which is now part of China. This was during the Japanese colonization of Manchuria. Renzankan is located at about the midpoint between what was then Hoten (present Hanyo), near Mukden (present Shenyang) and Anto (or Tanto, present Dandong) on the border with North Korea. At the time, there were occasional attacks by insurgents in the area, so he did not take us with him. In the fall of that same year, my father was shot in the leg during a patrol, and recuperated for two or three months at a military hospital before returning to duty.

Shimomura relatives seeing my father off upon his departure to Manchuria (1933), with me in the middle of the front row between my grandmothers.

My family moved from Fukuchiyama to Sasebo, Nagasaki Prefecture, immediately after my father's departure to Manchuria. Our new home in Sasebo was close to our relatives, the Sata family, in Shirahae-machi. Sasebo is a naval town and port with many sloping roads. Shirahae-machi is located on the slope rising from the Sasebo railroad station near the port up to Mt. Ebosidake. The exercise ground of the girls' high school was immediately below the house. At Sasebo, I attended the Seishin (Sacred Heart) Kindergarten near the railroad station for a year, while we lived together with my grandmother, Tsuki Shimomura. Her name means "moon," and she was a major influence on my early years. My grandmother and grandfather had lived for many years in the house of the Sata family, into which their daughter Shizue married, because their own house was taken as security for a loan.

1.2 Shimomura Family Traditions

My ancestors served the Nabeshima clan and were said also to be the headmen of Iko, Saigo-mura on the Shimabara Peninsula (presently Mizuho-machi, Unzen-shi); this peninsula is the site of the active volcano Mount Unzen, east of Nagasaki. My grandmother Tsuki, born in the late 1870s, was a direct descendant of the Shimomura family, while my grandfather Kosaburo was adopted. Traditionally, Japanese families that lacked sons adopted someone to carry on the family name, often a close relative or else a young man who married a daughter of the family.

Tsuki's uncle, Rinpachiro Shimomura (1831–1892), became acquainted with the industrialist Eiichi Shibusawa (1840–1931) at the Chiba *dojo*, a school of *kenjutsu* (fencing), in Edo, as Tokyo was then called. Eiichi Shibusawa, who visited Europe in 1867 and attended the world's fair, or the International Exposition, in Paris, was one of the most influential persons in the rapid industrial development of Japan in the Meiji era that initiated the rise of modern Japan. According to family traditions, Rinpachiro first became friends with Shibusawa's cousin Mr. Odaka at the Iba *dojo,* and later when Shibusawa commenced his sword lessons, Rinpachiro helped train him. Rinpachiro continued to be friends with him throughout his life.

In the 1880s, Shibusawa, by then a prominent financier and industrial entrepreneur, set out to build a modern factory to make bricks in the western way using German techniques (a Hoffman kiln) at his hometown, Fukaya, a city northwest of Tokyo that was already known for its ceramic tiles; Rinpachiro took on some job there. (His friend Mr. Odaka was also a Fukaya native.) Tokyo Station and other prominent historical buildings were constructed using the bricks manufactured in Fukaya. Rinpachiro worked there for about five years before he died at his house in Fukaya; his tomb is located at the Senkouji Temple there. The Shimomura house in Iko was probably built with financial help from Rinpachiro.

Since Rinpachiro did not have children, he had adopted someone from among the Shimomura relatives. This relative in turn adopted my father's youngest brother Tadashi as his successor. But my uncle Tadashi died young at age 24 in 1927, before I was born, and was buried at Rinpachiro's cemetery. My father had no chance to visit his brother's tomb after the war, and wanted to do so very much. He had planned to make the visit to Fukaya when I returned from my first trip to the United States, because it was located near Tokyo; unfortunately, he passed away before I returned.

We learned a few years ago, moreover, that during the war, the cemetery and the temple, which kept its documents, had been bombarded by B–29 attacks. The Senkouji Temple had been rebuilt since then, but many tombs of the cemetery were still missing or lost. The chief priest eventually found Rinpachiro's tombstone among the missing through search efforts by Mr. Fumihide Sawayama, my sister Sumiko's husband. However, Uncle Tadashi's tombstone remains missing.

In October of 2013, I finally visited Fukaya city with my wife and son to hold a memorial service for Rinpachiro's family in front of his tombstone in Senkouji Temple, as my father Chikara had most wanted to do. We met the mayor and city officials and they arranged a special sightseeing visit for us to the old brick factory, which had to stop operation in the 1980s when the construction of brick buildings was prohibited due to the risks posed by earthquake damage. During a quake, brick buildings are dangerously susceptible to collapsing, similar to what had occurred in California, as my son, who has lived in California, knew.

Shibusawa statue in Fukaya.

Shimomura house in Iko.

Since Fukaya is also famous for *Fukaya-negi* (green onions), they gifted me with a cute green-and-white stuffed animal representing its green-onion mascot; when my young granddaughters saw it, they were immediately delighted with it! They play with it still.

It appears that my grandfather Kosaburo dreamed of making a fortune in a single stroke. He borrowed money using the Shimomura house and premises as the security, but then his business went bankrupt, so he lost all his money as well as his home. Possibly because of these financial circumstances, his son Chikara—my father—had to attend the military academy, which was free, after completing middle school. So my father would have grown up in the shadow of his father's intense desire to remake his fortune and regain the Shimomura house. Mr. Eiichiro Sata, my grandparents' son-in-law, who was managing a small coalmine at the time, generously accommodated them in his own house. His business, incidentally, prospered during and after the war, and I recall that his house was quite roomy by the standards of the time, spacious enough to accommodate a governess in addition to his in-laws. Mr. Sata was also a very kind person. He often invited us for meals, and once he bought me an expensive mechanical toy set at the Tamaya department store.

The loan taken out by my grandfather eventually had to be paid back by my father and his younger brother Takeshi, so the loan also caused tensions between my parents. At the time, marriages were normally arranged by the parents of both families, who negotiated through an intermediary person. My mother, Yukie, was daughter to Yukimatu and Niwo Fujiyama; the Fujiyamas were a samurai family who owned land in Isahaya near Nagasaki. However, the existence of the loan was not made known to my mother's family while the match was being negotiated. Thus, when my mother found out about it after marriage, she must have been greatly shocked and upset, although I have no memories of it.

My grandfather Kosaburo died at age 69 in 1942 while still dwelling in the Sata house. Afterwards, Mr. Sata kindly made arrangements that resulted in the loan being completely paid off, so that the house at Iko finally became available to our family again a few years before the war ended. At that time, since Kosaburo had already passed away, Tsuki and Mr. Sata's daughter Mineko moved there together from Sasebo to take refuge from the bombing. I was probably fourteen then. The Iko house is registered as my permanent domicile in Japan, although I spent little of my childhood there and now have not lived in Japan for many decades.

1.3 In Manchuria

Some of my most vivid childhood memories are from Manchuria. Because the insurgents had been wiped out in the area, our family moved from Sasebo to Renzankan, Manchuria in March 1935. The trip took two days; we traveled by ship and railroad. In Renzankan, we lived on a quiet street of mostly Japanese residents in front of the railroad station, with official residences built of brick for army officers at the end of the street. My father was a major at the time and his residence, one side of a duplex, had three rooms. On the opposite side of the railroad station was a busy Manchurian town where various interesting foods such as steaming buns and candied red apples were sold, although I was seldom allowed to go there. In summer, there were enormous sunflower blooms, over one foot in size, in front of the entrances to the residences. A river flowed nearby and we could see the army barracks on the far side. Every morning, a soldier leading a horse used to come over a bridge downstream from us to escort my father to the barracks. The water level of the river increased drastically when the snow melted in the spring and also whenever we had heavy rain. At those times, the roaring, rushing water seemed to rise up above ground level, as if the river were about to burst over its banks.

I was enrolled at a Japanese school as a first-grade student together with three other children. The school must only have had two classrooms because my class included some third-grade students. It seems strange that I don't remember anything about my teacher, considering how very small my class was and that I stayed there for three whole years. Probably she taught me in a business-like, unmemorable manner, but since I had no problems in school after returning to Sasebo, she must have done her job effectively.

In winter, the exercise ground of the school was filled with water and turned into a skating rink. We used to order skates from Germany every autumn, and they would arrive via Siberian railroad by winter. It was a great pleasure to glide swiftly on the ice, but I hated having to sharpen the skate blades. The walk to school was really freezing cold when the temperature was –20 degrees Celsius. Our residence was heated from under the floor, and the inside was warm enough, except for one small room beside the entrance that was left unheated. We used this room as a freezer in winter, and I remember seeing several pheasants hanging there, probably gifts from someone.

After my mother's death, we found among her possessions this picture of my brother Sadamu and myself with my hat for kindergarten (March 1934).

I enjoyed fishing in the river in summer, and skating in winter, but there were no other pleasures over there. Every month I looked forward to the arrival of a magazine from Shogakkan (a publisher of magazines for school children), but it often did not arrive for unknown reasons. I always accompanied my younger brother Sadamu fishing, and we always caught plenty of small fish using earthworms as bait. My father liked to eat fried fish soaked in vinegar dressing. Every summer, Michiko, my uncle's oldest daughter, visited from Bujun to spend the summer vacation with us. (My uncle worked for the Manchurian railroad system and had six children. Together with other Japanese who were living in Manchuria, he lost everything after the Soviet invasion on August 8, 1945, so they shared the Shimomura house with us after they returned to Japan.) Michiko was one year younger than me and was a very good friend of mine.

Aunt Matsue, my mother's younger sister, visited us and went on sightseeing trips to Dairen (now Dalian) and Ryojun (Lüshun City). I had to accompany her to Shinkyo (Choshun), because it was unthinkable, according to ideas of morality at the time, for an unmarried young lady to travel unaccompanied.

There were three unforgettable incidents during those three years in Manchuria. The first of them happened when my sister Toshiko was six months old. One day, my mother took me, Sadamu, and Toshiko to Hoten (Shenyang) for some reason, and hired an open carriage for us. The road was very crowded with people so we moved very slowly. Suddenly, a man approached us and said, "I will buy that baby for one hundred yen." My mother instantly said that she would not sell her child. But the man stubbornly insisted and finally reached out his hands toward Toshiko. My mother became almost frantic as she sought to prevent him from seizing my sister. I didn't know what to do, but fortunately the man abruptly disappeared into the crowd. Incidentally, one hundred yen at that time would be worth three thousand dollars (300,000 yen) today.

The second incident was that Toshiko died of pneumonia in the middle of winter one year after her birth. There was no medicine for pneumonia at the time; however, my mother learned from a military doctor that wet mustard dressing might have some effect. Grasping at straws, she tried it, but to no avail. The effects of the mustard dressing proved harsh and cruel; Toshiko's small body swelled up and reddened. I remember Mother's deep sorrow after Toshiko's death.

The last episode involved a woman from Tottori, who had followed my father to Manchuria and had set up a residence in Anto (or Tanto, now Dandong). My mother found out about it, probably by some sixth sense, and there followed a series of stormy encounters between her and my father. Finally, Mother took a night train going south, taking Sadamu and me with her. The train arrived at Anto early on the following morning. Then Mother visited the woman at her house, taking us with her, and begged her to end the relationship with my father "for the sake of these children." The woman nodded, and then she served us breakfast: toasted *nori*, *miso* soup, and overnight pickles with rice. I remember that she ate *nori* after folding it into a tiny square. I thought, "This woman is a very fine lady," and I could not stop my feeling of sadness.

1.4 To Sasebo Again

Early in 1938, my father was transferred to an army position near the border with the Soviet Union. The rest of us moved back from Manchuria to Sasebo. I was enrolled in fourth grade at Shirahae Primary School. The school was located on a steep slope, so that I could see the submarines neatly parked in the port from the edge of the school's exercise field. My mother soon departed to Manchuria again to live with my father, while Sadamu and I stayed in the custody of our grandmother Tsuki for about one year until she returned. Because I was the elder son, Tsuki raised me carefully and strictly as the heir of the Shimomura family. To strengthen me physically, she tried to feed me with various nutritious foods like soybean milk and omelets filled with ground beef. She took me to a dentist, and upon finding a badly decayed molar, she ordered a gold crown for me. It took several months to finish the dental work, but I rather enjoyed going to the dentist because in the waiting room I could read various interesting storybooks, such as the popular serial about the crimefighter

Zenigata Heiji. I wanted to read more books after I finished my dental work, so I went to the city library often.

Grandmother was extremely strict in manners and etiquette, so I always had to maintain good posture in her presence. She liked to declare, "The samurai betrays no weakness when starving." After meals, she would check my bowl, and if she found even one grain of rice left in it, she would lecture me, saying, "Rice is produced by the hard work of peasants." I would then have to wash the inside of the bowl with hot water and drink it. After my bath, she would occasionally check behind my ears and on my neck. If she found any dirt, she would say that it would be ignominious when I had to be beheaded (because it is sometimes honorable for samurai to commit *harakiri* and then be beheaded). I knew she was emphasizing the importance of readiness, but her stern words were still a little scary.

Family picture in 1938 at Sasebo when my father returned home temporarily, with the Sata family. Front row from right: Sadamu, my younger brother; my father; Grandfather Kosaburo; Grandmother Tsuki; my mother Yukie. Back row from right: Osamu; cousin; my uncle Mr. Sata; cousin; cousin Mineko; governess Miss Kawahara; Mrs. Sata with her youngest son; cousin.

One day after my mother returned from Manchuria, we had a snowfall of about four inches. After morning exercises, we fifth-grade boys were ordered to take off our shoes and then had to walk barefoot in the snow for two hours around the foot of Mount Ebosidake. Fortunately, the sun started to shine and we did not feel the cold very much. I later heard, however, that the principal was scolded for that action by a city official.

My father's unit stopped at Moji while moving from Manchuria to Taiwan. Moji, a port city in Fukuoka prefecture, marks the narrowest point of the strait dividing the islands of Honshu and Kyushu, and prospered on international trade in the late nineteenth century. My mother took us two boys to Moji to meet with our father. At that time, Aunt Matsue also came with us and took care of me and Sadamu while Father was with Mother. Soon after, Mother visited Father in Taiwan for a month, since she knew she might never see him again due to the war in southern Asia.

During my fifth-grade year and the two years after, I spent my summer and New Year vacations with my mother's parents. My grandparents' house was in Nagano-machi at the edge of Isahaya city, and looked down on the plain of Isahaya under the foot of Mount Taradake. Grandfather was a leader of the neighborhood, and used to bring back a pack of delicious *momo* buns (peach buns, decorated with a pink picture on white buns) whenever he attended a meeting. Several tenant farmers who had worked for the Fujiyama family for many generations still lived nearby. Their main crops were rice and also trees grown to produce logs. Once I helped my grandfather to make charcoal. First we packed hardwood logs close together, covered them with earth, lit them on fire from one end, and left them for a few days to let the wood burn; then we closed off both ends. The charcoal was ready after one week.

A few days before New Year's Day, a few youths from among the tenants showed up and performed a dashing *mochitsuki*—a traditional pounding of rice to make rice cakes. In this ritual activity, three persons work together as a team. First, they put steamed sweet rice in a stone mortar, then two of them take turns pounding it rhythmically with a T-shaped wooden pounder while shouting. While they do this, the third person adjusts the shape of the *mochi* being pounded, using wet hands so that the rice paste would not stick to him. When the pounding was finished, I helped to make small round pieces of *mochi*. *Mochi* is customarily served to celebrate the New Year.

I often walked down the slope from the house and fished for small fish and crawfish in a nearby stream. Looking north, I could see the wide muddy shore of the Bay of Isahaya under the foot of Mount Taradake, where *mutsugoro* fish (mudskippers) are abundant. Although my grandparents liked to eat *mutsugoro*, I could not eat it because of its cute appearance. That muddy shore does not exist anymore due to a land reclamation project after the war. At some distance to the south, there was a beautiful seashore at Uki facing the Bay of Tachibana, and we used to go there for sea-bathing. Because the water was so clean then, we also made sea-salt by evaporating seawater in a large pan.

In April 1941, I entered Sasebo Middle School (secondary school), which was located in the northern part of the city. Japanese militarism had reached its peak at the time. We had to perform military exercises once or twice each week under direction of the officers attached to the school, and we had to wear gaiters below the knees. On one of these occasions, we first received a dress inspection, and we performed some routine exercises such as "Right, dress," "Right turn," "Go ahead," and "Halt." Then the military officer said, "You will fight for His Majesty the Emperor. Will you do *harakiri* if you make a mistake or fail?" Everybody replied "Yes." Then the officer said, "Take one step forward if you will do *harakiri*." Everybody else instantly took one step ahead, but I could not see the logic of the officer's words, so I did not go forward promptly. The officer looked at me and said, "What will you do?" Now I was prepared for punishment, and replied, "I would like to decide depending on the circumstances." The officer said, after a short pause, "You are very thoughtful. We have now finished today's exercises," and he left. I let out a sigh of relief.

In the second term of my first year at middle school, the school was relocated to a new place in distant Hiu-machi, south of the city.

Commuting by train was not permitted, so I had to walk. However, in December 1941, after the Japanese attack on Pearl Harbor, my family moved to Osaka, where my father was; thus I was at Sasebo Middle School only until the end of the second term of my first year. Like other people of my age, I graduated from middle school at the end of the fourth year, instead of the normal fifth year, without receiving a diploma.

But almost fifty years after our graduation, Mr. Takayoshi Shigenobu, a classmate of mine at Sasebo Middle School, tried to obtain diplomas for everyone in the class and succeeded. Mr. Shigenobu found out my mailing address in the United States and sent me the diploma, even though I studied at Sasebo Middle School for only eight or nine months. Incidentally, I never received a diploma from the middle school in Isahaya from which I actually graduated, but where I did not study for even one hour. All these irregularities were caused by the war.

Fifth grade class picture at Sasebo (1939), with me at left end, back row, because of my height. Beneath the small roofed hut is a photograph of the Emperor—the same as in every school, as part of Japanese militarism.

1.5 In Osaka

In the fall of 1941, my father was transferred from Taiwan to the Eighth Regiment in Osaka. He was a lieutenant colonel at the time. The Eighth Regiment was located near the Osaka castle, and well known by an old joke, "Once again defeated was Eighth Regiment," a joke probably originating from the nature of Osaka natives—Osaka people were generally fonder of comedy or sarcasm than Tokyo people, and the Eighth Regiment consisted mostly of Osaka natives. Children sing this phrase often, "Oh! Eighth Regiment lost again!" although the regiment actually had no particular record of defeat.

The Pacific War began on December 8th, and my family moved from Sasebo to Osaka. Our new home was in Takaai-machi, Higashisumiyoshi-ku, near the Tsurugaoka train station. By an arrangement made by a subordinate of my father, I was transferred to Sumiyoshi Middle School, which had a reputation as a bourgeois school.

Eighth grade at Sumiyoshi Middle School at Osaka (1942), with myself at the far left of the back row again, as the tallest in my class.

Thus, my middle school life in Osaka started with the third term of the first year. The teacher in charge of my class happened to be Shizuo Ito, a well-known poet from Isahaya. Its academic level clearly differed from Sasebo Middle School, however; my rank at the end of the term was a very poor 300th out of 302. In addition, I contracted tuberculosis quickly in Osaka, and felt very tired for about half a year, although I was allowed to attend school. I studied hard, but my rank did not improve much; I was only about 250th at the end of the second year. I was interested in making model airplanes at the time. Once I built a plane that flew by flapping its wings, based on an article in a scientific magazine, and displayed it at an exhibition at the Sogo department store. It flew about twenty meters horizontally, but could not ascend.

One day, my father took me to a special dining room of the Takashimaya department store to teach me table manners. He was in uniform for some reason unknown to me, although it did provide some advantages in that period, but I felt rather uncomfortable sitting in front of a uniformed person. The fried mackerel served had many small bones in it and I had great difficulty eating it with a fork and knife. Indeed, all kinds of food were in very short supply already at that time, and rice, the main staple of the Japanese diet, was rationed at about two-thirds of a pound per person per day. My sister Setsuko was born around this time (in 1943), so I had to do babysitting occasionally.

When I was in my third year of middle school, my father was sent to Thailand. The war turned clearly unfavorable to Japan. At school, missing lectures became common; however, I was glad when we missed military exercises. One day we were sent on an overnight march to Mount Ikoma. We started at 7:00 p.m. and, around midnight, we cooked rice in a mess tin and had a meal. We ate the cooked rice with herring that had been cut open and dried. We walked until 6 a.m., and then returned home by train. During the march, probably around 10 p.m., I noticed many tiny irregular spots of yellow light on the road. Now I think they must have been luminous earthworms crushed under our shoes.

By the time the second term began, our time spent in physical labor suddenly increased; we were put to building a reservoir to hold water. However, my rank at the end of my third year had improved to 160th. During my fourth year, my teachers appeared not to take their teaching duties seriously any longer. Mr. Ito preferred desultory talk to lectures, and he once said, "You are thinking that the Japanese military will soon bring out 10,000 Zero fighters from hiding to crush the enemy. Aren't you?" We all hoped that the military secret police had not heard him.

CHAPTER 2

The War Strikes Home: Ohmura Naval Aircraft Factory and the Atomic Bomb

During my fourth year in Sumiyoshi Middle School in 1944, my father wrote to tell us to move to the countryside to escape the American bombing. My mother chose to move us to the house of her parents in Isahaya. My grandparents gladly welcomed us. The house was on a hill looking down onto the plain of Isahaya. My grandfather had previously owned a considerable amount of land, which had been taken by the government to build an airfield. He still owned a small area of rice fields and some forests at that time, which were sufficient to support us. After the war, by American orders designed to reorganize Japanese society, many landlords also had to give away their land to their tenants, to allow them to be independent farmers. Therefore Mr. Fujiyama became poor after the war, but he still owned adequate farmland to live on. So he never worked for anybody. (After the Fujiyamas' son passed away at college age, they adopted my younger brother Sadamu, their grandson, as the family heir; while my aunt Matsue's younger son, my cousin Shigeru, was adopted by the Imazatos, my grandmother's maiden family.)

2.1 Student Mobilization Order at Isahaya Middle School

On September 1, 1944, I attended Isahaya Middle School for the first time, as it had been previously arranged. The school was located in the central part of the city, so it took an hour to get there from my grandparents' house. It was the first day of the second term of the fourth year. Because I was not acquainted with any of my classmates, I sat at the back of the room and waited. A teacher showed up soon and told us, "We have an important meeting. Please wait until I come back," then hastily disappeared. We continued to wait. After more than two hours, the teacher came back and told us, "In the teachers' meeting, we decided that this class will be mobilized to the Ohmura Naval Aircraft Factory immediately. Other classes will be mobilized to the Mitsubishi Shipyard and other arms factories in Nagasaki." Because the National General Mobilization Law had been activated recently, I expected to be mobilized sooner or later, but I hadn't thought it would happen on my first day at my new school. As a result, I never received even one hour of instruction at Isahaya Middle School.

At the Ohmura Naval Aircraft Factory, we lived in a wooden two-story dormitory. There were about twenty rooms, each occupied by six or seven students, with lavatories and a bathroom at one end of the ground floor. Our days started at 6:00 a.m. with physical training under the direction of a naval officer, followed by breakfast. We stayed there for less than two months before the factory was bombed and destroyed. Oddly enough, I don't remember what kind of work we did during that time, and I still wonder how they kept us all busy. The officer in charge probably had to go to great lengths to find odd jobs for us every day.

The main meal at the dormitory was cooked "rice." In the beginning, it was actually half rice and half wheat, but it soon changed to half rice and half defatted soybeans. Defatted soybeans were usually used as fertilizer and animal feed; to say the least, they were not tasty.

Moreover, because defatted soybeans are lighter than rice, the mixture of "rice" that we got sometimes consisted mostly of the defatted soybeans that made up the bulk of the volume. The side dishes were also poor in quality due to shortages of all kinds of food. I thought I would become ill from malnourishment if this diet continued for three months, but ironically, I was saved from that fate when the factory was completely destroyed by American bombing on October 25, 1944.

At about 10:00 a.m. on that day, the air-raid sirens sounded. Some students took shelter in an underground bunker. The rest of us ran to a ditch at the edge of the airfield. Then we waited. Soon I saw a formation of more than twenty American B–29s flying from the west, and the bombing started. It was the first time I ever saw B–29s and the blasts of their bombs. One stray bomb that fell nearby showered us with sand and gravel. The bombing was accurate and destroyed most of the facilities, but I saw that one hangar still stood. We noticed several people trying to pull a fighter plane out of it, so we rushed to help them, not knowing that an enemy plane was before us, invisible in the smoke-covered sky. Just as we reached the hangar, we were showered with numerous incendiary bombs. Someone shouted orders to take refuge. We ran between the bombs bursting into white flame; I saw someone who had received a direct hit on his shoulder running with one arm dangling. If those incendiary bombs had been the napalm that was used later in Vietnam, we would probably all have been burned alive.

When the air raid ended, we learned that some students had died in an underground bunker. Our dormitory had completely burned down. At around 4:00 p.m., we were ordered to return to our homes by train, and then ladies at an emergency kitchen provided us with *onigiri* (rice balls) of polished white rice, which would hardly have been available anywhere at the time. They were made from the scarce rice spared from the bombing attack. To us, that was the most delicious *onigiri* in the world.

Within one month after the Ohmura air raid, several wooden factory buildings were hurriedly built between the hills near Isahaya, and we were assigned to work there. Fortunately this factory was within walking distance from my home, so that I did not have to eat defatted soybeans anymore. It seems strange, however, that we sat idle for at least one month before the new factory began to function; we could have used that period for study at the school. Probably nobody thought about our schooling under such unsettled conditions, in which we couldn't predict our fate even for one day ahead. Therefore, I went to the factory every day even before it began to function. If, as frequently happened, I found nothing useful to do, I often lay down in a sweet-potato field nearby to watch the large formations of B–29s flying east, high above Mount Taradake. It was beautiful to see the B–29s shining silver against a background of blue autumn sky. About ten minutes after the bombers had passed by, we used to see black smoke in the Ohmuta industrial area located on the opposite shore of the Sea of Ariake, and we could only imagine the scenes of carnage over there.

The new factory was designed to repair the engines used in fighter planes (such as the Zero). When the factory became functional, I and a few other students were assigned to help four drafted workers in smoothing the face where the engine's crankcase joined the cylinder. Unlike us, the drafted workers had received technical training previously. They were all middle-aged men who ran small businesses, like a fish store or a barber shop, in distant places, and they were allowed to return home once every one or two months. We had a sufficient number of engines to repair for the first two or three months, but received very few after April 1945, probably because Japan had exhausted its planes in kamikaze attacks.

The duration of middle school education in Japan was originally five years, but the government shortened it to four years in that year for my class, so that young men could be sent onto the battlefield earlier. Thus, my class graduated in March 1945 at the factory without a diploma or even a graduation ceremony. Some who were accepted to the military school went there. The rest of us expected to be drafted to fight in the war soon, if the war did not end, except those who embarked on more advanced schooling. Although we had very little work to do at the factory, the student mobilization continued.

At about that same time, part of my grandfather's property was requisitioned by the naval paymaster's office. They said that their office was in an easily visible building in the airfield, so they had been ordered to move to the less visible hillside. After negotiation, we gave them an outlying building where my aunt Matsue, Sadamu and I had lived, and a storehouse. All our family lived together in the main house. Although we had to share our home with soldiers, they were always polite and never bothered us.

2.2 The Atomic Bomb on Nagasaki

Soon, the shadow of the war would loom even larger and darker in our lives. The news reported that on August 6, 1945, the city of Hiroshima was completely destroyed by a new type of bomb. We didn't know, however, that it was an atomic bomb. Then August 9 came. Shortly before 11 a.m. on that day, a siren sounded, signaling an air raid. Rather than going into a bunker, a couple of my friends and I went up to the top of a nearby hill as usual and watched the sky. I saw a single B–29 going from north to south toward Nagasaki, about ten miles away, and wondered because its course was unusual. The B–29 dropped two or three parachutes and I heard sporadic gunfire. I watched the parachutes carefully, but could not see any people hanging from them. Within minutes, another B–29 followed the first one on the same course, then a siren sounded the "all clear."

We returned to our factory building, expecting to resume work. As soon as I sat down on my work stool, a powerful flash of light hit us through the small windows. We were blinded and unable to see anything for about thirty seconds. Then, maybe forty seconds after the flash, we heard a loud sound and felt a sudden change of air pressure. We were sure that a huge explosion had occurred somewhere nearby, but we didn't yet know where.

The sky rapidly filled with dark clouds, and when I left the factory to return home, a drizzling rain had started; it was black rain, a mixture of ash, water, and nuclear fallout. When I arrived home an hour later, my white shirt had turned completely gray. My grandmother took one look at me and quickly readied a bath so that I could clean myself. That bath may have saved me from radiation poisoning caused by the fallout in the black rain.

The next morning, I noticed that one of the parachutes I had seen the day before was in the office of a technical officer. The officer said that the parachute had only carried measurement instruments and a transmitter. He also informed us that serious damage had occurred in Nagasaki, but no one knew the full details. The chief of the factory organized and sent out a rescue party. There were two ways to reach Nagasaki: one was by railroad, and the other was by a road that went over a mountain. Since the railroad had not been cleared yet, we were loaded onto several trucks that headed toward Nagasaki. We soon discovered, however, that the Himi Tunnel at the entrance to Nagasaki was impassable, and no side road leading to the city could be found. The chief of the rescue party decided to stop trying to reach Nagasaki. In the early afternoon, the railroad was opened as far as Michinoo, the second station from Nagasaki station, and injured people began to be rapidly transported to Isahaya and other cities nearby. I have no clear recollection of the next few days, although the miserable state of Nagasaki and Hiroshima must have been reported widely and the details made publicly known. Probably we were all in a state of mental shock.

The day of destiny came on August 15. Starting very early in the morning, the radio repeatedly announced, "Soon there will be a very important broadcast." It was to be the proclamation of unconditional surrender, broadcast by the Emperor Hirohito in his own voice. At noon, we and the people of the paymaster's office heard the broadcast together in our home. The signal quality of the broadcast was poor, but we clearly understood that the war had ended. We all heard the emperor's voice for the first time. Commonly referred to as the "Son of Heaven," he had never spoken publicly before, so his very voice reinforced the magnitude of the occasion. I think everybody breathed a sigh of relief, but also felt the fear of an uncertain future.

Several years passed before we learned the detailed information about the atomic bombs that struck Hiroshima and Nagasaki. Most importantly, the Nagasaki bomb was of a different type, far more powerful than the Hiroshima bomb. Even if the use of the Hiroshima bomb was justifiable on the grounds that it hastened the end of a debilitating war, the use of the Nagasaki bomb three days later cannot be justified. It was clearly a test of new arms.

2.3 After the War: Searching for a Future

The student mobilization ended automatically with the conclusion of the war, but I was concerned about my middle school record and what I would need to do to enter an upper school. In an attempt to receive some guidance, I visited the Isahaya Middle School one week after the end of the war. However, I realized that my thoughts were self-indulgent when I came to the entrance to the school. The school was surrounded by a long, tall iron fence, with its main gate to the right side. On the left, behind the fence, was a wide exercise field, and the school buildings were located behind the main gate. I saw several large sheets of paper attached to the side of the main gate; numerous names were written on them, about half of which were crossed out. The school was being used as accommodation for the injured people of Nagasaki, and their names were on the sheets. In the exercise field, several men were slowly strolling. They were all half-naked. The summer sun blazed down on them, and the air was filled with the noisy shrilling of the cicadas. Looking more closely, I saw that their upper bodies, mainly their backs, were covered with a black substance that had some white specks on it. At first, I thought they must have had some kind of black medicine applied to them; I later learned that it was dried blood. The white specks were maggots, larvae of flies that had hatched on their damaged flesh. However, the scene I saw in the next moment was even more shocking, and it may have affected me from that time forward.

In front of the gate, there was a cart parked on the opposite side of the road. On the cart were bodies covered with straw mats; one of them had its feet partially exposed. They were probably waiting to be taken to a crematory. Two people carrying a body on a stretcher were coming out of the school building. At that moment, I noticed two half-naked people standing near the fence, holding on to it with their hands. They had been watching the loading of the bodies. They had been observing the process that might be their own fate in a few days. I felt as if I was seeing ghosts. My brain froze, so that the shrilling of the cicadas faded. I thought all my senses had failed. I think the mental shock that I experienced in that moment, when I was just sixteen years old, has had a certain permanent effect on me. I have regretted for a long time that I did not speak with them; probably I lacked the courage to do so at that time.

It became clear by early September that the bombs used in Hiroshima and Nagasaki were atomic bombs. It was said that no plant would grow in the areas affected by the bombs. The injured victims of both cities were dying, one after another, every day. A rumor went around that the leaves of persimmon trees were effective in treating the bomb victims.

One day toward the end of September, while I was passing the days with nothing to do, I had a visitor at home. I could not recognize who he was until he told me his name, since more than half of his face was covered with a severe keloid. He said he was injured at the Mitsubishi Shipyard and afterward had been sent to a hospital in Nagasaki. He was one of the six persons whom I had shared a room with in the dormitory of the Ohmura Naval Aircraft Factory less than a year earlier. I remember that I had felt surprised at not seeing him when I was assigned to the Isahaya Factory. Since he was an evacuee from Korea or Manchuria and was living with his mother in Chijiwa on the Shimabara peninsula, he probably chose to be assigned instead to the

Mitsubishi Shipyard, which had a dormitory, rather than make the long commute to the Isahaya Factory.

He stayed one night at my home. The next morning, my mother prepared *onigiri*, balls of white rice, for his lunch, and then he departed for his home at Chijiwa. It was a hard night for me. I did not know how to talk with him; I tried to be cheerful but my heart was dark, considering that the bomb victims were dying every day. I have not heard anything about him since then.

A few months after the end of the war, I received about 800 yen as compensation for my labor during the student mobilization, which was unexpected because I had thought that we were doing unpaid service for our country. The inflation after the war was very severe, and the money and government bonds that my mother had saved to pay for education for my brother and me were now almost worthless.

I tried to enter Saga High School, Fukuoka High School, and Kumamoto Technical College in 1946 and 1947, but all three schools rejected me. In order to enter a high school or college at the time, a report on one's school record from middle school was very important. However, I could not expect a good report from the middle school that I had graduated from without having studied there for even one day. Therefore, I went back to the Isahaya Middle School to seek advice on what to do. Since I did not have a class teacher, I met the vice-principal. He considered my situation, but could not suggest a solution. I asked about the Wireless Communication Training School that was widely advertised. He rejected the idea, saying, "That is a place that will ruin a person like you." I asked him to give me another opportunity for advice in the future and left.

Half a year later, I heard that the Nagasaki pharmaceutical college destroyed by the atomic bomb was relocating its campus to the airfield that could be seen from my home. I met with the vice-principal again and asked about that school. He strongly recommended that I should apply there. I suppose he must have written a good report for me, because I was admitted to the Nagasaki Pharmacy College in April

1948. Although I was not interested in becoming a pharmacist, I saw no other choices under these circumstances. I had no other means to further my education. It must have been fate that opened this particular path to me.

2.4 After the War: Disappointments

On the other hand, at this time my father finally returned home from Thailand. As a lieutenant colonel, he had been ordered to negotiate with the British army when the Japanese troops in Thailand fell under its control. Then he made arrangements to ship about 20,000 Japanese soldiers back to Japan in an orderly way. He arrived in the Port of Yokosuga on the last troop transport ship. While he was supervising the soldiers unloading at the pier, someone stole all of his baggage. Thus he arrived home in Isahaya with nothing. Such occurrences were a part of life in Japan after the war.

Due to his military career, my father could not hold any official position or job after the war. So his traveling days were over; he stayed home. However, many people in the village respected him. Under other circumstances he might have become a leader in the community. He contributed to the local elementary school as a head PTA member for some years while my younger sisters were enrolled there. He also did various jobs around his property during this time, helped out on the Fujiyamas' rice field, and at one point raised chickens.

Since people did not then have widespread concern about many substances now known as hazardous, the farmers in the vicinity may not have taken precautions when handling agricultural chemicals. Exposure to chemicals and pesticides whose dangers were not known at the time may have eventually caused the pancreatic cancer from which Chikara died. We were all exposed to hazardous chemicals in those days. Akemi remembers how, on her school grounds, DDT was sprayed over their heads to kill insects in their hair several times after the war: "So we had to spend the whole day with white powder on our heads until we could go home and wash it away." This was on orders from the American occupational headquarters. No one understood the dangers.

Forty years after Chikara's death, on my birthday in August of 2002, Akemi typed up a note to leave our children about an autograph by the novelist Pearl S. Buck that we possessed. In the note, she explained that, in the fall of 1960, Pearl Buck visited Japan accompanied by a movie crew to film a movie based on one of her novels. At that time mainland China was closed to Americans, so her crew filmed in Japan instead. One location that they chose was Isahaya city, near Nagasaki. There, they used the Muta family residence, just a few blocks away from my parents' house in Nagano-machi, in the Ono District; they did some of the filming in a field owned by Muta-san, who was a considerable landowner.

When Pearl Buck heard that there was a family nearby whose son had been studying in the United States—that is, she heard about me while I was away on the Fulbright—she came to meet my parents and talked at length with my father. She learned that I had gone to Princeton University. Before departing from my parents' house, she autographed two pages, one for my father and one to be given to me.

At that time, she was residing in Bucks County, Pennsylvania, which was just over the state line from New Jersey. She extended my father a warm invitation for me to visit her sometime while I was in Princeton, because she lived close by. She wrote her address and phone number on the back of the paper. I found the autograph that was kept in my father's house after my mother passed away in 1993. It was visibly worn by thirty years in the humid Japanese climate, when compared to the one that I had held in my keeping all those years.

I never found the opportunity to visit Pearl Buck. Originally it was because I was uncertain about going to visit someone so famous during my first stay in Princeton. Later, when I was back in Princeton, I had a young family, and we never found the right time to go. My father passed away in 1962, only two years after meeting Pearl Buck, while Akemi and I were still living in Princeton. We regretted losing the chance to hear more about her visit to his home.

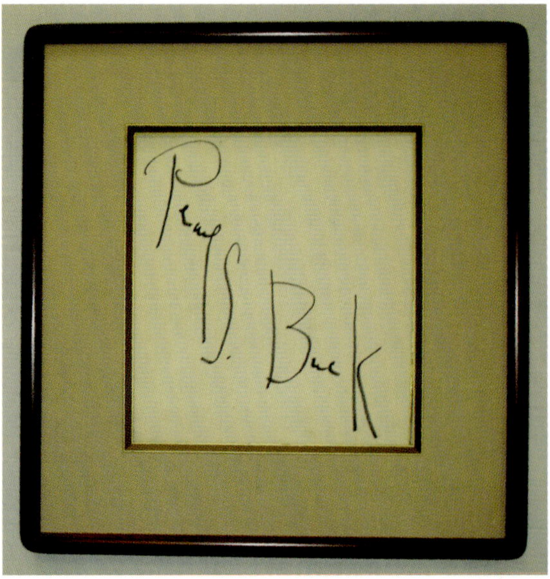

Pearl S. Buck's signature given to Chikara during her 1960 visit in Isahaya.

Akemi and her brother at my Fujiyama grandparents' cemetery, Isahaya.

CHAPTER 3

Struggling to Learn: Nagasaki, Nagoya, and Luciferin (1948–57)

Nagasaki Pharmacy College belonged to the Nagasaki Medical College, and laid claim to a long history beginning with the arrival of Philipp Franz von Siebold, a Dutch physician, in 1824. Both colleges were completely destroyed by the atomic bomb in August 1945.

3.1 Nagasaki Pharmacy College and Nagasaki University

The surviving professors and students of the Pharmacy College opened a temporary campus in 1946 in Saga; then the campus was relocated to Isahaya in 1948, using the buildings of the former Airplane Pilot School. At the time I entered the Pharmacy School, the campus consisted of several wooden buildings. Despite the great efforts put forth by the professors, the equipment was still poor. Although we had no problem receiving lectures, we lacked resources to perform experiments easily. Professor Takatori of Pharmaceutical Botany was a world-class scholar, but the training level of the other professors was much lower than it should have been, due in large part to the long-lasting war. The preparation of the students was inferior as well because their studies had been so thoroughly interrupted by the war. While I received adequate instruction in analytical chemistry and physical chemistry, I learned much less about organic chemistry. There were mandatory German lessons instead of English for the first two years; I suppose it was because they could not find an English teacher. However, most students showed no interest at all in German, so I felt sorry for the teacher. My college expenses were paid with my government-sponsored loan.

Nagasaki Pharmacy College, temporary campus (1948).

My grandmother, who was living with us, presented me with a new set of student clothes when I entered the college. There were almost no clothes available for purchase at the time, because of the economic state of Japan after the war, but she owned a mulberry field. So she raised silkworms, reeled the silk thread off the cocoons, dyed it and hand-wove it into cloth, and even tailored it into a men's jacket for me. The suit was too luxurious for everyday use, and I also felt a little hesitant to wear it to campus, although everybody admired the jacket.

Most students, who lived in the dormitory, were constantly hungry due to an extreme shortage of food supplies following the war. I lived at my family home, which included some farmland, so I was fortunate with respect to food. Since we had little opportunity for recreation in such a war-torn countryside, many people enjoyed competing at Go and playing the guitar. I played tennis occasionally. While performing organic synthesis experiments, we often inadvertently ignited solvent fires due to poor technique and poor glassware. But we had no difficulty in putting them out, because many of us were seasoned firefighters from our wartime experiences. There was a pile of blankets in one corner of the room that was left over from the former Airplane Pilot School. We would take one of them, wet it with water, and then cover the source of the fire with it. Ether and alcohol fires could be put out leaving no trace, but benzene fires would leave black soot all over the room; therefore, we were most careful when using benzene.

With my classmates, wearing the silk suit made by my grandmother.

I became increasingly interested in chemical experiments, but the experiments I could freely do were practically limited to inorganic ionic reactions, considering the equipment available. I asked permission of Professor Yasunaga of the Pharmaceutical Analysis Lab to prepare many glass capillary tubes of an inside diameter of 1 millimeter. These were packed with alumina powder before a very small amount of a mixed solution of metallic ions was applied to the tip of the capillary, which was then dipped into a developing reagent. As the developing reagent quickly rose up into the tube by capillary action, the metallic ions were separated, creating colored bands corresponding to different types of ions. This process was a type of chromatography. I gained further permission from Professor Yasunaga to bring home small amounts of various chemicals in order to study the conditions of these separations in detail. A few years later, in 1953, we reported the results in the *Journal of the Pharmaceutical Society of Japan.* This was my first paper.

In the laboratory, Nagasaki Pharmacy College (myself: fourth from left).

I graduated from the Nagasaki Pharmacy School, first in class rank, in March 1951. As a result of the reorganization of the Japanese educational system, the Pharmacy School was converted into the Department of Pharmacy, Nagasaki University. I was offered a transfer into the new university system as a fourth-year student, but I declined the offer because it meant the repetition of a year. No further education was immediately available to me. I applied instead for a research position at Takeda Pharmaceutical Company. After an interview, however, one of the interviewers kindly advised me that my personality would not fit in with a company position. I am now sure that he was correct, although I lacked the experience to know it at the time. Professor Yasunaga offered me a job as an assistant in the analytical chemistry laboratory for students, and I took it.

The new Department of Pharmacy was located in Showa-machi, about a mile and a half north of Urakami where its predecessor had been located before the bomb was dropped. It was a three-story reinforced concrete building formerly used as a teachers' school, with a bend in its central portion that was caused by the atomic bomb blast. There were a number of changes in the new educational system, and a doctoral degree became mandatory for professors. A few professors lacked a doctoral degree at that time, including Professor Yasunaga, although he soon obtained the degree through work undertaken at Kyoto University.

In my job in the laboratory for students, I collaborated with another assistant, Mr. Yamaguchi (Akashi), who was adopted later and therefore had two family names. In my spare time, I continued the chromatography experiments that I had started as a student. I also developed an interest in classical music after I had listened to an LP record somewhere; however, one LP record cost 3,000 yen at the time, while my salary was only 4,000 yen. Somehow I managed to buy a record player and a speaker, then assembled an amplifier from glass vacuum tubes, condensers, and resisters, probably with some financial help from my parents.

I lived in the city of Nagasaki for a total of seven years, renting rooms in various places. The best and longest accommodation I had was with the Maekawas, a devout Christian family, in a second-floor room of a detached house. They provided morning and evening meals for me, as well as a bath, and they called a doctor for me when I developed acute appendicitis. They were really kind-hearted people.

Professor Shungo Yasunaga was a very gentle and kind person, and well trusted by those who knew him. Four years after I started working for him, he obtained permission for me to take a paid leave of absence to study for one year. He even offered to introduce me to Professor Fujio Egami of Nagoya University, a well-known molecular biologist. I was greatly surprised, so I asked Professor Yasunaga, "Do you know him well?" He replied, "I met him only once at a conference. But there is no problem because he is from Yamaguchi prefecture, like me." I remember that I was greatly impressed by the strong connection between Yamaguchi people.

Professor Yasunaga took me to Nagoya by an express train (*unzen*) that left Nagasaki at 7:00 in the evening. Upon arrival in Nagoya the next morning, we visited Professor Egami's office, but unfortunately he was absent. Professor Yasunaga visited another person from Yamaguchi, Professor Yoshimasa Hirata, taking me with him. At Professor Hirata's office, Professor Yasunaga explained why we were in Nagoya and mentioned that Professor Egami was absent; then we chatted for a few minutes. When we were leaving, however, Professor Hirata said to me, "Please come to my lab. You may start at any time." Professor Yasunaga was embarrassed. As soon as we were outside the door, he asked me, "What will you do?" At that time, Professor Hirata was a natural products chemist who had become a full professor one year before.

Shungo Yasunaga (1911–1959).

Yoshimasa Hirata (1915–2000).

I had no preference for either professor's field of expertise. Since I knew nothing about either molecular biology or natural products chemistry, one choice seemed as good as the other. I thought that Professor Hirata's offer might be the direction given me by heaven, so I accepted his offer of a place in the Hirata lab. It seems to me that this decision determined my future by ensuring that my life work would be the study of bioluminescence, aequorin and GFP.

To my great regret, Professor Yasunaga died of cancer only four years later at the age of forty-eight.

3.2 The Hirata Lab and *Cypridina* Luciferin

In April 1955, I enrolled as a research student in the Hirata laboratory, in the Department of Science at Nagoya University. I found a room in Mr. Ohara's house, a ten minutes' walk from the lab, which provided me with two meals and a bath. Both sides of the road to the lab were overgrown with wild azalea. The Department of Science was in a three-story concrete building, and the Hirata lab was on the first floor. The lab consisted of a large laboratory room that accommodated about twenty people, another room for large-scale experimental work, and two small rooms. I was assigned an experimental table near the entrance of the large laboratory room. Near my table were the tables of Mr. Hattori and Mr. Nakano, both of whom were research students. Mr. Hattori was sent from Nagoya Sanken (a government research center), and Mr. Nakano was a former naval officer sent from the Fujisawa pharmaceutical company. The experimental table of Professor Hirata was by the window near the center of the room, surrounded by the tables of the assistants, Mr. Goto and Mr. Kakizawa, Mr. Mamoru Ohashi, and about ten students. Professor Hirata used to come earlier than anyone and was always doing experiments except during lecture and conference periods.

On my first day at the Hirata lab, Professor Hirata produced a large vacuum desiccator from somewhere and told me, "This contains dried *Cypridina* (sea firefly)." They were crustaceans each about the size of a sesame seed. Then he explained how *Cypridina* emits light through the interaction of an organic compound called luciferin and an enzyme luciferase, that the luciferin is extremely unstable and decomposes in the presence of oxygen, and that Professor Newton Harvey of Princeton University had been trying to purify the luciferin for the past twenty years but had not successfully obtained a pure luciferin suitable for structural study. Harvey visited Japan in 1916 during his honeymoon and concluded that *Cypridina* was a good research material for the study of bioluminescence. Since then, he had continued his study using material obtained from Japan.

Then Professor Hirata asked me, "Could you purify and crystallize *Cypridina* luciferin for the purpose of structure determination?" At the time, crystallization was the only practical method to confirm that a substance was pure. Professor Hirata also explained, "I cannot give this subject to a student who is pursuing a degree because of its uncertain outcome." I understood the difficulty of the task. However, I was in

Nagoya for study and not for a degree, so I immediately replied, "I would like to do my best."

Specially made Soxhlet apparatus.

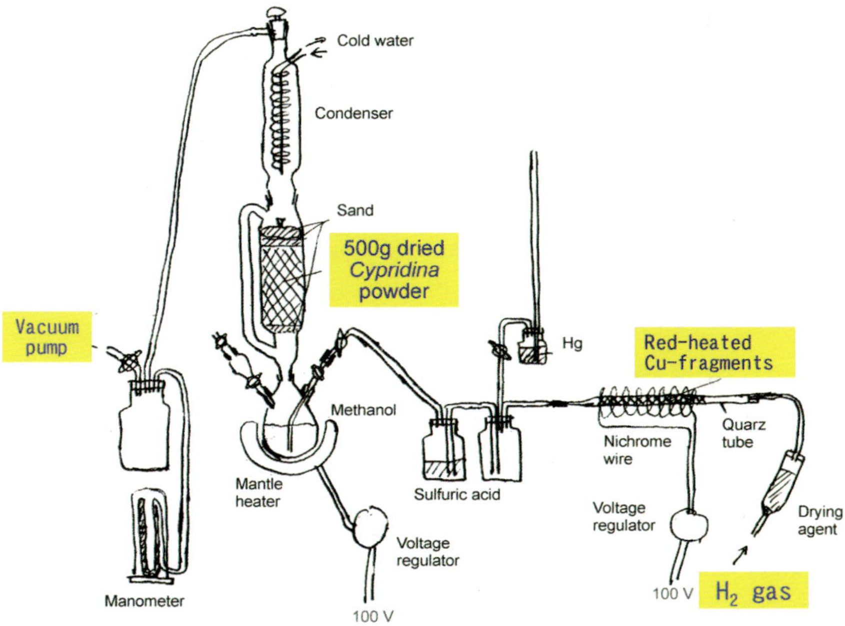

Left: live glowing *Cypridina*, photographed by us shortly after collection. Right: dried *Cypridina*, collected in 1944 and kept dry, still gave off light in 2009 when I ground it up and added water. Credit: Tom Kleindinst.

First, I checked the status of the progress that had been made at Princeton. I found that researchers there had devised a creative and effective purification method of the luciferin in 1935, and that recently they had succeeded in measuring its absorption spectrum. According to the Anderson method, water-soluble luciferin is converted into a fat-soluble form by benzoylation, and water-soluble impurities are washed out, then the fat-soluble luciferin is converted back into water-soluble form by hydrolysis. I still don't know how Anderson could have found such a method without any knowledge of the luciferin's structure.

In the first month of my work, I concentrated on obtaining preliminary data for the extraction and purification of luciferin. I found that methanol was a good solvent for the extraction, that cellulose powder and alumina could be used for the purification, and that these experiments must be performed in hydrogen gas or argon to prevent oxidation of the luciferin. So I would need an apparatus set up to remove air from the container in which the experimental steps would occur.

I judged that at least 2–3 milligrams of purified luciferin would be needed for the crystallization. To obtain such an amount of purified luciferin, I estimated that we would need to extract 500 grams of dried *Cypridina*, ten times the maximum amount they used at Princeton. I reported these matters to Professor Hirata, and he immediately ordered a glass-blower to make a giant-sized Soxhlet apparatus (see diagram).

Then I was ready to begin the real experiment.

It was a complicated and lengthy process. After I assembled the extraction apparatus in the large workroom, I ground 500 grams of dried *Cypridina* into coarse powder, using a mortar, and packed it into the Soxhlet. The *Cypridina* powder was first defatted with benzene, and the defatted powder was dried with an aspirator. Then, the air inside of the apparatus was replaced with purified hydrogen gas which had been passed through first a quartz tube heated red-hot, containing copper fragments, and then a wash bottle containing concentrated sulfuric acid in order to remove the oxygen. (While going through the quartz tube, any oxygen within the hydrogen gas would change to water; therefore sulfuric acid, which has dehydrating properties, was used to remove this newly made water inside of the hydrogen gas.) The purified hydrogen gas bubbled through the methanol into the extraction flask of the Soxhlet.

When the air inside of the apparatus was completely replaced with hydrogen, the stopcock of the hydrogen inlet to the flask was closed and hydrogen gas inside the apparatus was evacuated—pulled out— from the top of the Soxhlet to lower the inside pressure. (Low pressures make it easy to boil methanol at a low temperature.) So I had to adjust how much gas to evacuate, watching until the methanol started boiling, and also adjust the temperature. Then, when the methanol in the flask started to boil, the extraction process started. When the temperature of the flask was adjusted to about 45 degrees Celsius and the stopcock at the top of the apparatus was closed, boiling and extraction continued for many hours without further adjustment.

Although hydrogen is not dangerous when carefully used, I recall that nobody came near me. I chose to use hydrogen due to the ease of removing any trace of oxygen from it; it is very difficult to remove trace oxygen from nitrogen and argon.

When the extraction was finished, the extract in the flask was concentrated using an aspirator and purified first by the Anderson method, followed by chromatography on cellulose powder or alumina. This allowed me to further separate out the luciferin. This purified luciferin was used in the crystallization experiments. The absorption spectrum at this stage was very similar with that of the best sample obtained at Princeton. Absorption spectrums at the time were measured with a Beckman DU spectrometer, which was a tedious process requiring measurement at every wavelength. Fortunately, however, Professor Hirata instructed my friend Goto to begin to help me in my work. The extraction and purification had progressed smoothly up to this point, and all went according to my plans, but the crystallization was not as easy.

At Hirata lab, 1956.

I tried to crystallize purified luciferin using combinations of various solvents and salts. Examination under a microscope showed only amorphous precipitates, however, without a trace of crystals. In addition, the oxidation and decomposition occurred faster than I expected, even when I took the utmost care, and the material became unusable in two days. The process of the extraction, purification and crystallization trials took about one week of almost sleepless work. To make further attempts at crystallization, I had to purify a new luciferin sample from dried *Cypridina*. The samples whose crystallization had failed and those that had oxidized were used in the isolation of oxidation products, elementary analysis, and amino acid analysis, in the hope of obtaining some information on the structure of luciferin.

3.3 An Unexpected Success: Luciferin Crystals

I continued to extract and purify luciferin about once a month, because I could not obtain any crystals, no matter how hard I tried. However, the luciferin was finally crystallized by accident on a cold day in February 1956.

The night before, I was trying to crystallize the luciferin as usual, but I had exhausted all my ideas by about 10:00, although there was a small amount of purified luciferin left. I decided to use it in amino acid analysis. I added an equal amount of concentrated hydrochloric acid to the luciferin sample. The yellow color of the sample instantly changed to a dark red. Since an oven was not ready, I decided to heat the sample the next morning. I left the dark red solution on a shelf and went home. When I came back in the morning, the solution was colorless. I thought that the sample was hydrolyzed, but upon close inspection I found a tiny amount of black precipitate at the bottom of the test tube. Under the microscope, this precipitate could be seen as red needle-shaped crystals.

After successful crystallization, with Toshio Goto at Nagoya (1956).

The crystals were confirmed to be luciferin by luminescence reaction with luciferase. Because luciferin was thought to be a peptide-type compound based on previous amino acid analyses, its crystallization with strong acid was entirely unexpected. The decrease in room temperature overnight might have helped crystallization, because the laboratory was heated with a gas stove that was turned off at night.

My success in crystallizing the luciferin was accidental; however, I had accomplished what the Princeton researchers couldn't, and it gave me hope for the future, which had seemed dark since the end of the war. I was so happy that I couldn't sleep for a couple of days. However, the most important reward I received was probably the self-confidence I gained, that any complex problem could eventually be solved by effort.

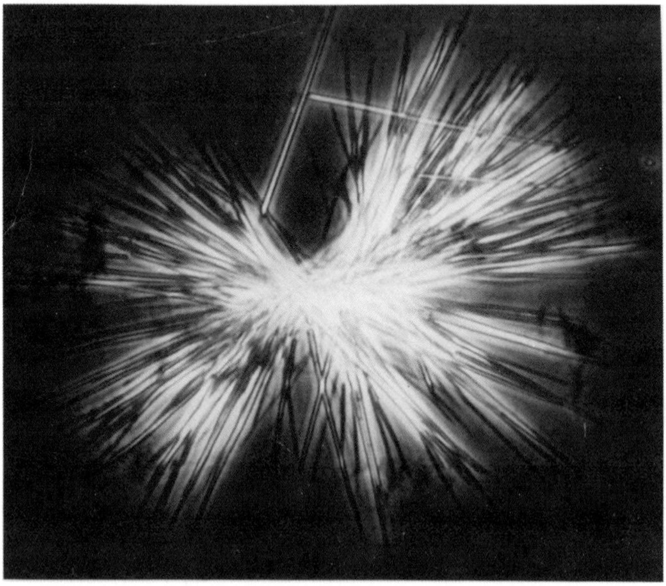

***Cypridina* crystals; their actual color was a deep red.**

As a consequence of the crystallization, the possibility of determining the structure of *Cypridina* luciferin was now open, and so Professor Hirata made a request to Nagasaki University to extend my stay in Nagoya by one year. My work on the purification of the luciferin became much easier now because the properties of the pure luciferin were already known. I discovered that the luciferin molecule contains the residues of the amino acids tryptophan, isoleucine, and arginine. The crystallization of luciferin in a strong acid like hydrochloric acid can be explained by the presence of an arginine residue that contains a strong alkaline guanidine group. (Generally a strong acid and alkaline combination has a tendency to make stable salts like the commonly known compounds NaCl, NaOH and HCl.) The chromophore structure of the luciferin still remained to be discovered. (The chromophore is a chemical group that produces the color that will be visible in the luminescence reaction.)

In 1957, our first paper on *Cypridina* luciferin was published. We later learned that in the summer of the same year Dr. Johnson of Princeton University, a former student of Harvey, visited Japan and extracted *Cypridina* luciferin from the live organisms for the first time at Shimoda Marine Laboratory. Until that time, only dried *Cypridina* had been used in the study of luciferin. Because it was found that drying of live *Cypridina*, usually under sunlight, causes a large loss of luciferin by oxidation, we used only fresh *Cypridina* that had been frozen with dry ice for the extraction of luciferin thereafter.

A memorable experience from my Nagoya years is the five-day mountain climb along the Yari–Hotaka traverse in the Northern Japan Alps in 1957. My friends Goto, Matsuura, Mizuno and I departed from the Nagoya railroad station around midnight and arrived in Kamikochi before noon the next day. We immediately started climbing, and slept in tents for the night. Then we tracked across Mount Jonen, Mount Yari, and Mount Hotaka. The route was not easy for a beginner like me, but I enjoyed the thrilling experience and the beautiful scenery.

It is fortunate that much of my research in later years would lead me to destinations of natural beauty around the world; numerous bioluminescent creatures frequent the scenic mountains, caves, and seas of the earth.

CHAPTER 4

To America and Back: Marriage, Princeton, and Discovery of Aequorin and GFP

Professor Frank H. Johnson sent me a letter, in the spring of 1959, inviting me to come work at his laboratory at Princeton University. I accepted his offer and promised to arrive the following summer, but I declined his offer of travel expenses. Instead, I applied for a Fulbright grant. It turned out to be the right decision because the Fulbright grant provided me with various experiences that would otherwise have been unavailable to me. I reported my plan to go to Princeton to Professor Hirata. Then he suggested that I should receive a doctoral degree, adding that I might have a salary of $6,000 instead of $3,000 if I were to have a doctorate, and he also offered to handle all the paperwork himself. Professor Hirata's offer was something that I had never imagined, since I was not at the time enrolled in a doctoral program. I accepted his offer with thanks.

I recall that the Hirata lab was a wonderful laboratory maintaining a splendid atmosphere. Nobody taught me anything formally there, but I learned many things by observing other people and was then able to develop various techniques myself. Only once did Professor Hirata tell me, "Mr. Shimomura. Mass spectrum can tell you an exact molecular weight, not the approximate one," but I no longer remember the circumstances under which he instructed me about that.

Incidentally, the scientific study of bioluminescence had often crossed the divide between Japan and the United States. Professor E. Newton Harvey of Princeton University, who pioneered the study of bioluminescence and who counted Dr. Frank H. Johnson among his students, first studied luminescent ostracods (tiny crustaceans, smaller and more rounded than shrimp) in Japan a century ago, in 1916. He even stocked the ostracod *Cypridina* from Japan in his lab, as I was to discover later. Dr. Johnson collaborated with Dr. Haneda of Yokosuka City Museum when he visited Japan in 1957, and they found that the luciferin and luciferase of the luminous fish, *Parapriacanthus*, would cross-react with the luciferin and luciferase of *Cypridina*. Namely, *Parapriacanthus* luciferin emits light when *Cypridina* luciferase is added, and *Cypridina* luciferin emits light when *Parapriacanthus* luciferase is added. Dr. Johnson thought that *Parapriacanthus* luciferin and *Cypridina* luciferin were chemically identical. To prove his hypothesis, Dr. Johnson visited Japan again in 1960, accompanied by his wife. Dr. Haneda collected the fish for the experiment, and they planned to perform the experiment at Professor Sugiyama's lab in Tokyo Kyoiku University (presently University of Tsukuba, a major research institution). Professor Sugiyama called and asked me to join them for the experiment and, when I agreed, he reserved a hotel room for me. So I went to Tokyo. The hotel reserved for me was the same one where the Johnsons were staying; thus I met Dr. Johnson there for the first time.

Late next morning, I went to Professor Sugiyama's lab together with the Johnsons to make the preparations for the experiment. Dr. Haneda and Miss Saiga were also there. I remember that it was a hot and humid day and Professor Sugiyama offered a small bottle of cold milk to each person. I accepted it with thanks, but the Johnsons declined the offer. Probably Professor Sugiyama did not realize that milk is a food, not a refresher drink, in the western world. In Japan at the time, milk was a precious and luxurious drink. The most popular refresher drink in Japan back then was probably Bireley's orange juice, an orange-flavored sweetened water.

Regarding the experiment, there already were isolated pyloric caeca obtained from over one thousand fish by dissection. The pyloric caeca, adjacent to the stomach of the fish, seems to function as part of its digestive system, and had been determined to contain (and possibly store) luciferin. We extracted the pyloric caeca with cold methanol the next morning. The content of luciferin was so high in that organ that we could easily crystallize the luciferin after chromatography. Dr. Johnson successfully proved the luciferins of the two species to be identical.

4.1 Marriage with Akemi: Departure from Japan

On August 27, 1960, I left Yokohama on the ocean liner *Hikawa-maru* for Seattle with over 200 Fulbright fellow scholars and students. It was my thirty-second birthday. I had just gotten married. Because it was also the last Pacific cruise of *Hikawa-maru*, the pier was completely filled with people. (*Hikawa-maru* had made her first trip to Seattle thirty years earlier, and she had weathered the war as a hospital ship, while her sister-ships had been destroyed.) My new wife, her mother, and some of my friends were among the crowd. There were thousands of festive colored tapes between people on the boat and the people on the pier, connecting them. I will never forget the vivid scene as the ship began to move and numerous tapes started to break and then fall down toward the water.

Departure of Hikawa-Maru, Yokohama (1960).

My wife Akemi and I had married very rapidly before I left, and she was to join me as soon as she had received her visa. Our marriage was arranged in the traditional manner. I thought I needed a partner to live with me while I was in the United States for a few years but had no idea about whom to ask. My aunt, Mrs. Matsue Takeda, also thought that I should take a bride with me because my stay would be a couple of years long.

It was then only three months before my voyage. So she hurriedly looked around for a potential candidate, inquiring among her friends for a suitable young lady. Mrs. Baba, who was a former neighbor and friend to her, had also been a classmate of Akemi's mother, Mrs. Okubo, at their middle school. Mrs. Baba thought Akemi might be a good match and thus she acted as an intermediary between the Takeda and Okubo families, introducing each family to the other. Mrs. Okubo found out that my father had been one of her brother's friends in middle school. In addition, in the town where she had been born, some of her relatives knew my grandmother Tsuki very well, because Tsuki's mother had been born in that same town. The result was that she was well disposed toward the possible match.

Half a month later my aunt showed me the name Akemi Okubo and asked me for my opinion. I was very surprised to hear Akemi's name because she had graduated from the same department of Nagasaki University as I had, and one year previously she had even attended a short course that I had organized. I remembered her well because she had a pearl ring on her finger, and I had to advise her to be careful of chemicals that might damage the pearl. And I had gotten a rather good impression of her. I replied to my aunt that she was all right with me, and to please go ahead with the marriage arrangements. But my aunt recommended that I see her again before making a final decision, and so we met a few times during that month.

Akemi was not one hundred percent sure about this marriage (as was typical for arranged marriages of the time, where the proposed couple rarely knew each other well), but she agreed to it, hoping that her choice would turn out to be a good one. She thought my grandmother was just like her own mother, and this helped bring about her decision. Her mother was reluctant to send Akemi to a faraway country, at a time when very few Japanese travelled abroad. Akemi, however, was curious to see Princeton, even though she was first surprised to hear that I would be going there, Princeton being home to many scientists associated with the Manhattan Project. She felt some special destiny guided her to a connection with the atomic bomb, which had so affected her childhood. Her father was completely satisfied with her decision, but her mother always wanted her back in Japan, closer to her.

Akemi was descended from the Honda family, and she lived in the Honda family house with her aunt and uncle during the war. This house was known for containing a special lacquered bathroom, as well as a special gate and entrance hall made for the lord's palanquin. These were reserved for the Nabeshima clan lords who had maintained a friendship with the Honda family over nearly fifteen generations, and would come visit regularly. Akemi and her cousins were not allowed to use that bathroom or entry, of course. The gate was always closed except when it was opened up for cleaning.

The Hondas were not a samurai family but were descended from retainers of the Nabeshima clan; a cadet branch of that family had property around the Shimabara peninsula. According to her family's traditions, around four hundred years ago the Honda family head had met and married the daughter of the overlord of the Nabeshimas, Takanobu of the Ryuzoji clan, who had been killed in battle in 1584. The Nabeshima were a large clan of Hizen province, now Saga Prefecture, next to Nagasaki. The Ryuzoji had controlled a larger territory that included part of Nagasaki, and including Shimabara peninsula. So the Nabeshimas had eventually inherited their overlord's lands, and since he had no surviving son, his daughter had sought protection for herself from her father's loyal supporter, the head of the Nabeshima clan, who was then living in Kojiro, on the Shimabara peninsula. When she married into the Honda family, the Nabeshimas maintained a friendship with her and her descendants thereafter. Akemi's family roots thus lay not far away from my own; we were both descended from retainers of the Nabeshima clan and raised by a mother or grandmother with strict traditional beliefs.

Less than a month remained before my voyage by the time both families had agreed and finalized our engagement. I was busy with preparations for my departure and could not spare any time for a celebration such as a wedding party. We had a wedding ceremony at Tokyo on the fourth of August, witnessed by my aunt, Mrs. Takeda, and Professor Goro Kobayashi, who was the chairman of the Pharmaceutical Department of Nagasaki University and staying that summer in Tokyo. Professor and Mrs. Kobayashi made every arrangement for us, finding both the place for the ceremony and a clergyman. They selected a fairly small and quiet location for the wedding. I think they themselves had held their wedding there several years ago. (That building no longer exists now, after the big development of Tokyo; I am very sorry for this.) The ceremony was western-style rather than Buddhist. We heard later that the clergyman was an uncle of one of the Kobayashis, since the clergyman whom they had originally scheduled had unexpectedly become unable to attend.

Our wedding in Tokyo, with Professor and Mrs. Kobayashi; my aunt Mrs. Takeda and the clergyman are in back (1960).

The next day, in response to strong urgings from my father, I flew back to Nagasaki with Akemi for a little lunch reception with our relatives. At the time, I was reluctant to do so, but this opportunity turned out to be the last time I was able to see my father. He seemed to be very satisfied with my marriage and in a very good mood.

During the thirteen days of the journey to Seattle, the sea was calm, the meals were good, and it was a comfortable voyage. However, there was one unusual happening. It was probably the seventh day of the voyage and we were near the Aleutians. About a half hour after I had gone to bed, the deck became noisy. I woke up and went up onto the deck. I learned that someone had fallen into the sea, and a crewman had thrown a lifesaving buoy, but it was impossible to stop the ship quickly. The person who fell was a graduate student whose life must have been full of hope and promise. The ship circled in vain for four hours to find him. The water temperature was 12 degrees Celsius, and at that temperature a person cannot survive more than thirty minutes. It was nearly a full moon on that night, and the sea was very calm with almost no ripples. I thought, if it was not a suicide, that he might have been bewitched to step off into the sea by the lovely moon, because it was so beautiful that night.

We saw Vancouver Island on our left side, then the next morning we arrived in Seattle within a dense fog. We had three days of orientation at the University of Washington, but it was mostly observation tours and sightseeing. I ate half of a cantaloupe for the first time in my life and I liked it. (Melons are not rare in Japan but still very expensive.) Pink grapefruit was also very good; for years I would eat half a grapefruit at breakfast each day whenever available.

We travelled from Seattle to Chicago on the Northern Pacific railroad. It took two days on a Pullman car with a private room. I ate meals on the dining car and enjoyed the scenery from a second-floor observation car in the daytime. Such luxury was made possible probably because the United States was then enjoying a golden age, and the Fulbright program was managed directly by the State Department. I stayed in Chicago for one night, then traveled to

Philadelphia, again by Pullman car, and finally arrived in Princeton. It was my twenty-first day after leaving Yokohama. Incidentally, after the retirement of the ocean liner *Hikawa-maru* in 1960, all Japanese Fulbright grant recipients traveled by air. The journey would have been considerably shorter, but much less full of memorable experiences.

The 1960 Fulbright Scholars aboard Hikawa-maru.

The Fulbright fellows aboard the *Hikawa-maru* included good friends of mine from Nagoya University, Dr. Reiji Okazaki and his wife Dr. Tsuneko Okazaki. On board we often enjoyed *sukiyaki* parties. *Sukiyaki* is a traditional Japanese hot-pot dish that is prepared in a pot on the middle of the table. The Okazakis were going to Washington University in St. Louis, so we took the same train from Seattle to Chicago, and I enjoyed the scenery and the experience with them. They are best known for their important work on DNA replication and the discovery of the Okazaki Fragments, whose understanding has led to further research on genetic mutations and cancer. Reiji was much missed after his early death, due to leukemia, as a result of his exposure to radiation from the atomic bomb in Hiroshima. He was just 44 years old, and if he had lived longer, he would surely have won a Nobel Prize, as many people still maintain. His wife Tsuneko is still alive; we met her at Nagoya in 2009, when we reminisced about our journey together so many years ago, and she gave us some pictures of that occasion. The mathematician Dr. Mikio Sato, laureate of the Wolf Prize (2003), who was going to Princeton's Institute for Advanced Study, was also aboard the same ship.

4.2 To Princeton

I arrived in Princeton on September 17, 1960, and spent my first night at Dr. Johnson's house. The next morning, we started to look for an apartment for me. We visited a house which we saw in the advertisement section of a newspaper, and rang the doorbell. A man came out and opened the door, but he shut it again as soon as he saw my face. It was clearly a racially motivated response. Dr. Johnson apologized to me with overwhelming sincerity. I experienced similar racial discrimination two or three times during the next year, but have not encountered such problems since then, at least on the surface. In any case, I found a good apartment through a real estate agent about ten minutes' walk away from the lab. It was the second floor of Mr. and Mrs. Philip T. Carroll's house, located near the corner of Nassau and Harrison Streets. The apartment had three fairly large rooms, a Victorian bath and a kitchen, with an independent entrance. The rent was $100 per month. I heard that this nearly 200-year-old house at 338 Nassau Street was purchased by the famous architect Michael Graves two years after Mrs. Carroll passed away in 1987.

I remembered the Carrolls fondly through the years, though we did not keep in touch. In 2008, immediately following the announcement of my Nobel Prize, we were extremely touched to receive a thoughtful email letter sent to the MBL from the Carrolls' son Tom (Thomas); it brought back many memories to us. As he recalled in the letter, the Carrolls had experienced strong opposition from their neighbor for renting to us. Sometimes threatening phone calls came too—including an anonymous death threat. Actually, I had even felt the unusual tension and had asked Mrs. Carroll what was happening. But she had replied, "You do not need to worry, please stay here as long as you like."

Once Akemi joined me in Princeton, Tom, who was then in his early teens, often helped her learn English in the afternoons, teaching her the pronunciations for items in the living room, the currency, and the foods at the neighborhood grocery store after he came home from school. He remembered her as "a very eager and entertaining learner." Akemi, who was used to being surrounded by a large family back in Japan, including four younger brothers and two sisters, was glad to have such congenial company. Later, Tom and his mother would help Akemi catch fireflies in the backyard in the summer evenings for my research.

In the dining room of our apartment with my favorite drinks. (Scotch was too expensive for me to afford back in Japan).

To show our appreciation for the Carroll family's kindness, we folded and strung strings of a thousand multicolored origami cranes for a Christmas gift to them before we moved out. Akemi collected food wrappers and other pretty everyday paper to cut into origami squares. Traditionally, a thousand cranes symbolize good luck, good health and long life, or a wish. Like dragons, cranes held a special place in Japanese folk beliefs, and then came to be widely associated with the wish for peace, in memory of victims, especially children, of the atomic bombs at Hiroshima and Nagasaki. Tom told us in his letter that he still had these cranes, which they used to put up on the door to the living room each Christmas season, spread out in the shape of a Christmas tree. He wrote:

> I didn't know then what a significant symbol of peace that was for her to have made for us, but I learned the story of Sadako Sasaki and the thousand peace cranes a little later in my life, and I treasure that string of cranes as one of my most important possessions. We've had it mounted in a shadow box, and it hangs on the wall in the main hall of our house here in Troy, New York. A year or two ago, the local Children's Museum of Science and Technology had a "cranes project" in which they taught local children to fold origami cranes and then had them fold a thousand of them to be hung from strings on a frame as a work of art. [...] I was invited to speak at the Children's Museum while they were in the middle of that project. I brought Akemi's cranes with me and told them the story of these two lovely people from Nagasaki who were an important part of my childhood. The woman at the Children's Museum then taught me how to fold a crane.

Tom also remembered that we had hung a picture of a molecule on the wall of our apartment, probably luciferin. He later went on to attend Caltech, and worked a while at JPL as my son Tsutomu also would, a decade or so afterward. It seems our families remained in some way connected even when we did not know it. I feel that some good destiny brought us to meet such warmhearted Americans as the Carrolls, the Johnsons, and later the Tuttles who befriended Tsutomu.

Dr. Johnson's room at the university was in the brand-new Moffett Laboratory that was attached to the left side of the old Guyot Hall. When we came in the main entrance of Guyot Hall, there were displays of a huge skeleton of a dinosaur and preserved samples of plants and animals. In later years, my children enjoyed exploring the exhibits on Sundays.

When I first visited Dr. Johnson's room, he produced a small vial filled with white powder, explaining, "This is the freeze-dried light organs of the luminous jellyfish *Aequorea*, and it should emit light when mixed with water." We went into a dark room to test it, but we could not observe any light. However, he enthusiastically described to me how abundant *Aequorea* were in the sea at Friday Harbor, in Washington state, and described how brilliantly luminous they were. Then he asked me if I was interested in studying its bioluminescence. I didn't know anything about the jellyfish, but was eager to study a new luminous organism, so I answered, "I will be glad to do it." Thus, we agreed to travel to Friday Harbor the following June.

Dr. Johnson (photo lit by luminescent bacteria).

Guyot Hall with newly built attached Moffett Laboratory (Johnson's lab was upstairs at the right end of photo).

In my first few months at Princeton, I worked to purify and crystallize the crude *Cypridina* luciferin that Dr. Johnson had extracted from live *Cypridina* in 1959 in Shimoda, Japan. Then I extracted and purified *Cypridina* luciferin from dried *Cypridina*. The Biology department stocked surprisingly large quantities of dried *Cypridina*. The supply included a large amount that Professor Harvey (who had died in 1959) had obtained from Japan, some in sealed bottles dated 1928 (my birth year). I saw one sealed tin can bearing the name of Sakyo Kanda, who was a pioneer in studying bioluminescence. In addition, there were large amounts that had been collected during the war in Japan by military orders; its intended use was as a dim light marker within tropical jungles. These supplies were seized by the US Navy at the end of the war and eventually donated to Professor Harvey's lab.

Miss Yo Saiga, who had helped me a few months before at Professor Sugiyama's lab, arrived from Tokyo in December of 1960, to work as our assistant. Akemi arrived the following month on a day of heavy snow, just after the inauguration of President John F. Kennedy.

At the time, a physical examination and proof of financial support were mandatory to enter the United States, and a delay in obtaining her visa had caused my wife's late arrival.

It took Akemi a little time to adapt to being in America. Generally Japanese people tended to look young to American eyes, so once a shoe repairman thought Akemi was a fourteen-year-old girl, a decade younger than her actual age, and gave her a lollipop. Another time, when she was driving, a policeman stopped her because she looked to him to be too young to drive; by then, she may have been close to thirty. (She needed a cushion under her and behind her to be able to see well out the windshield of our big American car, though, since she was only about five feet tall.) I remembered this when, many years later, my niece Satoko (my wife's elder sister's daughter) visited us, and because she was also very short by American standards, everyone treated her as though she were a child, although she was already in her twenties at the time. Many Japanese men also had to show identification to purchase liquor at the store. My father's family is exceptionally tall by Japanese norms, though, so I never experienced such difficulties.

Akemi in front of the Carrolls' house on Nassau St.

4.3 Friday Harbor

On June 23, 1961, Dr. Johnson, Miss Saiga, and my wife and I set off for Friday Harbor on the west coast, as we had planned. Dr. Johnson purchased a new Plymouth station wagon for the trip. We loaded his car with a large photometer (to measure light emission), other research instruments and chemicals that we would need, and even fit suitcases for four people upon the roof. The floor of the car became considerably lower due to the weight of the baggage. We took the Pennsylvania Turnpike to Chicago, then went north to take US-2, and thereafter we drove to Anacortes, Washington, mostly on US-2, since there were yet no interstate highway systems like there are now. Dr. Johnson drove all the way alone, for at least ten hours every day: he was a tough driver.

During this trip, there was a long drive in desert-like land under the strong summer sun in North Dakota, and we all suffered a little from the heat and dryness, because the car lacked an air conditioner. However, in the next state, Montana, we took a wonderful side trip to Glacier National Park. There, we stayed one night in a cabin at Many Glacier, where we saw a hole near the top of the Rocky Mountains soaring to the west. When the sun went down in the evening, we saw its light through that hole. On the next day, we went over the Rocky Mountains taking the "Going to the Sun" road. The rocky peaks and glaciers created astonishing scenic views. On the following day, in the eastern portions of Washington State, we drove through another desert-like place for a long time, then at the foot of the Cascade Mountains we bought and ate some large black Bing cherries. They were juicy, sweet and delicious, and I cannot forget how good they tasted. In many future trips to Friday Harbor, we would buy big boxes of Bing cherries, and, later, my children would cherish this tradition. Akemi would make cherry pies that she would freeze, and we would eat them throughout the year until they were gone. The Cascade Mountains were filled with greenery, very different from the Rocky

Mountains, which was refreshing after driving through the desert-like places. We went down the mountains, and we were finally on the west coast.

At Glacier National Park in Montana, the day before we reached the Continental Divide, on the trip to Friday Harbor (1961).

At the ferry terminal in Anacortes, we waited in the line of cars and fortunately we were able to take the first ferryboat. Two hours later we arrived in the small town of Friday Harbor. It took us seven days to cross the continent, driving about 3,000 miles. We drove a couple of miles from the port of Friday Harbor to the Friday Harbor Laboratories of the University of Washington, and met with the director, Dr. Fernald. Akemi and I stayed in a dormitory room for the night, but the room had no shower. Since I had back pain caused by the hard seat of the Plymouth as well as the bad road at that time, Dr. Fernald kindly arranged for us to have an apartment in the Tourist Hotel in town. The apartment had a large room with a kitchen and a bathroom and was quite a satisfactory place for the summer, although it had a minor flaw: the floor was slightly sloped and everything round would roll down to one side of the room. The Hotel was managed by an old couple who had escaped communist Yugoslavia and who were very kind to us. It took about fifteen minutes to walk to the lab, but Mr. Hironobu Ozaki, who was a graduate student of the University of Washington and lived nearby, often gave me a ride.

Friday Harbor is on San Juan Island, which is about ten miles long and located east of the city of Victoria on Vancouver Island, Canada. The town of Friday Harbor at the time was concentrated near the ferry dock. It had only several stores, a small movie theater that opened on weekends, and a bar, which was the only social spot on the island. There were two historical places, American Camp and English Camp, near each end of the island, both remnants of the Pig War (1859). The Pig War had originated in a border dispute between America and English Canada. After a pig owned by a Canadian had eaten vegetables on a field owned by an American, the American shot and killed the pig, which triggered the establishment of military camps by both sides on the island. However, the dispute ended later without turning into a real war. English Camp is located near Roche Harbor, an old historical resort, where we could see the remains of a kiln where limestone had been burned many years ago. I saw an old tombstone of a Japanese man who was probably an immigrant laborer who worked at the kiln in the Meiji era.

On the island at the time, rabbits propagated profusely, and the roads were scattered with numerous bodies of rabbits hit by cars; they were later picked up by eagles. Near the laboratories, at a place called Heaven Hill, at the end of a hike up a zigzag path, we could see bald eagles soaring. I remember the master of the hotel kindly offered us raw rabbit meat, hunted by his son, but my wife did not know how to cook it. The cooking stove was an old-style big iron box, whose surface heated pots and pans, and Akemi had trouble with this unfamiliar way of cooking.

We enjoyed the splendid scenery of the island. The sea was clean and beautiful then. The fish were abundant. Fishing from boats or from the rocky coast, we easily caught rockfish over a foot long, using just simple lures from a boat or from the rocky coast. We enjoyed them as sashimi. At low tide, we could see sea urchins and starfish of various sizes and hues scattered across the rocks, along with occasional abalones, which we considered a delicacy. It was tricky to pry an abalone off a rock, though. If the first attempt failed, the abalone would stick tightly on the rock, and become impossible to dislodge. We were not permitted to take more than two abalones per day per person by Washington State law.

Later, when we returned to Friday Harbor each summer with our children, it became a family ritual to go fishing at our favorite spot on the rocky black coast nearly every weekend. It was Akemi who eventually caught the largest fish, a yard-long lingcod. When she was fishing from the shore, a fish bit her lure. She wound her reel and brought up the fish near the surface of the sea. At that moment a big lingcod showed up and bit the fish with its huge, wide-open mouth. Akemi fought with that lingcod for nearly ten minutes, and then eventually pulled it near the shore. The fish was finally brought out of the water using a hand net, with help from Arthur, Mr. Ozaki's son, who was fishing nearby. My children were very excited that their mother had caught such a big fish, and they told and retold that story for years. Indeed, Friday Harbor in those years was a kind of paradise for my family.

Another dear remembrance at Friday Harbor is about Miss Scott. When Akemi and I were walking a country road during that first summer, she stopped and offered us a ride. After that, she gave us occasional rides and took us sightseeing on the island on Sundays. She was a Public Health Service nurse, who had been responsible for the previous twenty-five years for an area of about 400 square miles where close to 3,000 people lived on several islands. If a case of chicken-pox or a childbirth took place, she would rush to the spot by car, ferry, water plane, or any available means. She was near retirement, but filled with great spirits and energy.

One day, Miss Scott took us to Roche Harbor, a famous resort that even President Roosevelt had visited once. We saw the remains of the kiln, the marina, a famous old hotel with a pretty garden, and the mausoleum of the founder of the Roche Harbor community. Everyone except the tourists knew Miss Scott. There was a small store isolated from the marina, and near the store Miss Scott found a blond girl of fourteen years or younger wearing a bikini, with two boys. Miss Scott introduced us to the girl: "This is a girl I took care of at her birth." At the time bikinis were not very common yet. The girl was very shy. However, when I had noticed some blond hair coming out of her bikini by chance, I was troubled about where to look. She must have been a very innocent girl. Miss Scott probably also noticed my embarrassment and said, "Let's go on to the next place," and we departed.

Then we visited the English Camp. There was a small two-story log cabin on a broad field near a beach. The walls of the second floor of the cabin had small holes on all sides, probably to extrude the barrel of a rifle. I supposed that a rifle shot at the time could not penetrate a log. Then, we drove through a road high above a cliff called Dead Man's Bay and returned to Friday Harbor. I later found that there are several places named Dead Man's Bay in North America, all under steep cliffs.

On a different day, Miss Scott accompanied us to nearby Orcas Island, which took thirty minutes by ferry, when she had to go there for an errand. She showed us many places on the island, and then we took dinner at the Rosario Resort. It was a gorgeous resort of a type rarely seen in such a countryside location; it was previously the mansion of the Moran family who made a fortune by shipbuilding. The restaurant was in a luxurious building which had a built-in pipe organ. We inadvertently spent too much time at dinner, so that Miss Scott had to hurry to the ferry dock nearly ten miles away, driving at a speed of nearly 75 miles per hour on a country road, and we caught the ferry.

At Friday Harbor Laboratories, 1961 (Mt. Rainier and Mt. Baker could be seen in the distance on clear days).

A cluster of the jellyfish *Aequorea* in the water.

4.4 The Jellyfish *Aequorea*

For our research at the Friday Harbor Labs, Director Fernald assigned us a part of one large laboratory room, shared with three other scientists. One of them was Dr. Dixy Lee Ray, a professor at the University of Washington at the time, and later Commissioner of the US Atomic Energy Commission and Governor of Washington state. She was always accompanied by a dog, her well-known trademark. Although dogs were prohibited in the Friday Harbor Laboratory by state law (because of its role as a biological preserve), she declared that the animal was her assistant, not a dog. She once collaborated with Dr. Johnson in research. Another scientist there was the marine biologist Dr. Eugene Kozroff, who was still working in 2004 when I visited FHL.

The *Aequorea* were extremely abundant. We would see a stream of floating jellyfish sweeping by alongside the lab dock in the early mornings and evenings, riding upon currents caused by the tides. We carefully scooped up the larger ones into buckets, one by one, using a shallow dip-net to avoid damaging them. A specimen of *Aequorea* is formed like a hemispherical umbrella, nearly transparent, with faint radiating lines along the outer half. An average specimen measures 7–8 centimeters in diameter and weighs about 50 grams.

Aequorea, when stimulated, emits light along the edge of the umbrella, displaying a green ring in darkness. Because its luminous organs are located only along the edge, we cut off the margin of the umbrella, making a strip of 2–3 millimeters in width, and we called this a ring. We used the rings in our experiments. By squeezing about thirty rings through gauze, we obtained a turbid liquid which emitted weak light for a long time.

At the time, everybody believed the luciferin-luciferase hypothesis, which presumed that all bioluminescence reactions were caused by the reaction of a substrate called luciferin and an enzyme luciferase, as in the cases of the fireflies and *Cypridina*. Raphaël Dubois had discovered the luciferin-luciferase reaction in extracts from a luminescent West

Indies beetle in 1885, and then found a corresponding result with the luminous clam *Pholas*. Newton Harvey built on that work and published the first substantial book on bioluminescence, as he termed luminescence in living organisms. Firefly luciferin had been crystallized within about a year of when I had succeeded in crystallizing *Cypridina* luciferin. As an example, the firefly luciferin reaction is catalyzed by luciferase in a two-step process. In the presence of the luciferase, the firefly luciferin reacts with adenosine triphosphate (ATP), a substance vital to living cells, to produce luciferyl adenylate and pyrophosphate; the luciferyl adenylate then oxidizes (reacts with oxygen) quickly, emitting a cold yellow-green light. This is the source of the firefly's light.

So we attempted to isolate a luciferin and a luciferase from *Aequorea*. We tried every conceivable method, but they all failed. So we ran out of ideas after working for only a few days. I became convinced that our dependence on the luciferin-luciferase hypothesis was causing our failure. Therefore, I urged Dr. Johnson to stop focusing on extracting luciferin and luciferase and, instead, to try separating out the luminescent substance whatever it might be. He was, however, unwilling to abandon the hypothesis. Since we disagreed so intensely on our experimental procedures, I started working to one side of the table, while on the other side, Dr. Johnson and his assistant Yo Saiga continued their unavailing attempts. The situation was awkward; we all felt uncomfortable.

It would be an obvious premise that, in order to extract a luminescent substance in its active form, light emission must be prevented during extraction; if light emission were to take place during extraction, the extracted substance would be "used up" and would no longer emit light. Because luciferin emits light through an enzyme-catalyzed oxidation, I tried to extract the luminescent substance of *Aequorea* using various enzyme inhibitors and oxidation inhibitors. I tried all the inhibitors that I could think of, but none of them worked, and I exhausted my ideas after ten days.

I gave up on doing experiments, and tried to imagine the reactions that might occur in luminescing jellyfish, searching for a way to inhibit the bioluminescence reaction. I thought hard day and night. I have a bad habit of not speaking to or answering my wife at home once I am deep in thought, but she always fed me without a single complaint.

Often, I meditated on a drifting rowboat under the clear summer sky, so that nobody would disturb me. There was little traffic at sea around Friday Harbor at the time, and since a rowboat has the right of way over any vessel with a motor, ferries always kept a wide berth. However, if I fell asleep and the tidal currents carried the boat away, I needed to row for a long time to return to the dock. One afternoon on the boat, a thought suddenly came to me. It was quite a simple idea: *Even if luciferin and luciferase are not involved in the jellyfish bioluminescence, some protein is probably involved in the bioluminescence reaction. If so, the activity of that protein would very likely be altered or affected by a change of pH.* Immediately, I returned to the lab, made buffer solutions at various pH values, and extracted the rings using these solutions. Extracts at pH 5, 6, and 7 were luminous, but the extract at pH 4 was dark, suggesting that the acidity had inhibited the luminescence reaction. I filtered the pH 4 extract to remove cell debris and tissues. The filtrate became luminous when it was neutralized with sodium bicarbonate, indicating that the luminescent substance of *Aequorea* could be extracted at pH 4. I thought we had finally found the right method to extract the *Aequorea* bioluminescence substance.

However, I was surprised when I threw away the luminescing solution into a sink. The inside of the sink lit up instantly with bright blue light! Because there was seawater flowing into the sink, I suspected that seawater had triggered the reaction. Since the composition of seawater is well known, I quickly determined that calcium ions had caused the sudden luminescence. It was sheer luck that some seawater was in the sink at the right time. I reported my findings to Dr. Johnson. He was very surprised at first, and then delighted. I do not remember that we celebrated in any way, but I think Dr. Johnson might have gone to the bar with his friend Dr. Dixy Ray. After that occasion, I enjoyed his full confidence in research matters.

Thus, we found that the bioluminescence of *Aequorea* was caused through the action of calcium ions. It was already known that calcium ions could be bound by the calcium chelator EDTA (which was known at that time as Versene), and we found that using EDTA would inhibit the luminescence of the *Aequorea* bioluminescence substance more effectively than using a pH 4 solution. Therefore, we decided to extract as many jellyfish as possible utilizing EDTA. We could collect the jellyfish easily, and we could also buy them from the children who lived by the laboratory at one cent apiece. But, there was still the problem of cutting off the rings. The jellyfish was slippery, and cutting a narrow width from the margin of its umbrella using a pair of scissors required considerable skill, and it took at least a minute for each jellyfish. We hired ten high school girls and paid two cents apiece, but this plan did not work well. However, with the efforts of everyone in our team, we were able to extract about 10,000 jellyfish by the end of August. Then Dr. and Mrs. Johnson packed the extract with dry ice and brought it back to Princeton by car. Mrs. Johnson was a librarian at the Chemistry Library of Princeton University, and she had arrived in Friday Harbor only two weeks earlier.

My wife and I, together with Miss Saiga, returned to Princeton by railroad. We took the Northern Pacific train that I had taken the previous year from Seattle to Livingstone, Montana, and we all enjoyed sightseeing at the Yellowstone National Park for four days. I remember that we had a heavy snow on September 3 at the Yellowstone National Park. Then, we went to Chicago and saw the old city and the Art Institute of Chicago, and finally arrived back in Princeton.

4.5 The Busy Year of 1962: Finding the First Traces of GFP

In September of 1961, we started purification of the crude bioluminescence substance obtained at Friday Harbor. This crude material obtained from 10,000 jellyfish contained about 50 grams of protein. Thus we needed to perform the first step, column chromatography, four times, using a 3.5 centimeter diameter column. The only suitable adsorbent available for column chromatography was DEAE cellulose, so we repeated the DEAE chromatography, varying the solvent pH at each step. In February of 1962, we had obtained about 5 milligrams of nearly pure luminescent substance. Since the substance was a protein, we named it "aequorin" based on the name of the jellyfish. Aequorin attracted wide attention as a unique protein that would emit light in the presence of calcium ions, even in the absence of oxygen. Aequorin was the first discovered photoprotein, and various other photoproteins were discovered later.

During the column chromatography of aequorin, we found a trace amount of protein that showed green fluorescence and eluted sooner than aequorin, and we also purified it. The protein is now called green fluorescent protein, or GFP, and its fame and applications have far exceeded those of aequorin.

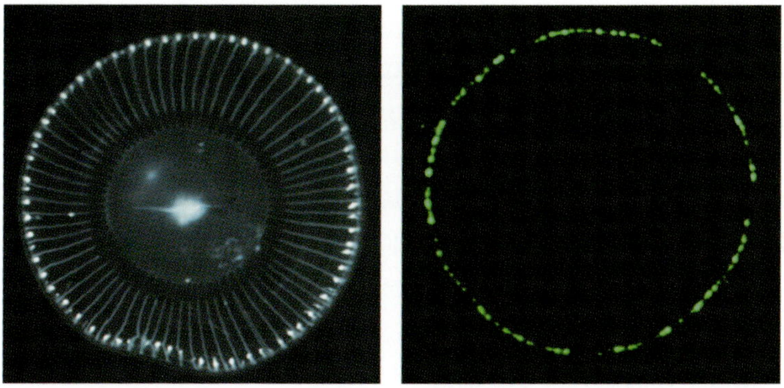

Left: single *Aequorea* in ocean; right: luminescent *Aequorea* in dark room (aequorin glows bluish, but living jellyfish glow greenish because of GFP).

In April of 1962, my wife and I attended a Fulbright conference held in Washington, DC. After the conference we were given a tour of various places including Mount Vernon where George Washington had lived, the Library of Congress, and the National Museum of Art. At the White House, we could not meet President Kennedy due to his need to attend to urgent business at the time (it was during the lead-up to the Cuban Missile crisis), but instead we met his younger brother, Attorney General Robert Kennedy, at the State Department. He was waiting at the entrance of a room, and he greeted and shook hands with each of us. I felt that he was a great person. (Senator Fulbright also could not attend for the same reason, and Mrs. Fulbright made a speech instead.) It is regrettable that, following the assassination of President Kennedy a year and half later, Democratic Presidential candidate Robert Kennedy was also assassinated six years later and I had to see off his funeral train at the station of Princeton Junction.

We returned from Washington to Princeton, then immediately flew to Phoenix, Arizona, accompanying Miss Saiga, to see the famous Grand Canyon. Then, we flew to Los Angeles to do preliminary study on the sea worm *Chaetopterus* and the yellow flower-like sea anemone or coral *Parazoanthus* at Dr. Rimmon Fay's Lab at the University of Southern California. Dr. Fay was an experienced diver, and he frequently helped us obtain research specimens thereafter. *Chaetopterus*, also known as parchment tube worm, is a feathery segmented worm that creates a curved tube to dwell in, formed of a thin tough substance that feels somewhat like parchment or rice paper; we had to cut it out of its tube with scissors. In Los Angeles, we met Dr. Nobusuke Kawano of the Pharmacy Department of Nagasaki University, who was visiting UCLA.

On May 10, 1962, we flew to Seattle, then on to Friday Harbor on a small six-seater plane. There, we collected the jellyfish *Aequorea* and extracted aequorin from them just as we had done in the previous year. We did not get so many jellyfish this time, probably due to the early season, but we flew back from Seattle to Newark, located north of Princeton, because we had other trips on our schedule for the summer. For that flight, we chose a seven-stop plane that would fly lower so that we could see the scenery better, although the airline offered us a non-stop flight at the same fare; we were still young and energetic. However, I felt strangely sad when I saw the setting sun from the plane. I later learned that my father had died in Japan on that day.

4.6 Fireworms in Bermuda

On August 10, 1962, we flew to Bermuda to study the fireworm *Odontosyllis,* a segmented sea-dwelling worm that employs luminescence in its mating and reproduction. The female worm's green light attracts male worms to her. This creature may have been the source of the light display seen by Christopher Columbus's expedition in 1492, and interpreted by the sailors as an indication of land, as they voyaged through the Bahamas. We had a much easier trip to see the fireworms, a two-hour flight from New York. Our party consisted of Dr. Johnson, his daughter Frances, Miss Saiga, and my wife and me. At Customs upon arrival, I was asked, "Do you belong to NASA?" and I was confused because I had never heard of NASA, although everybody knows it now. I learned later that NASA was established only one or two years before, and there was a NASA station in Bermuda. Bermuda is an independent country in the Atlantic, but it was formerly a British colony and the old monetary system was still in use at the time, with twenty shillings to the pound and five pence to the shilling. This system confused my wife in her everyday shopping. The marine laboratory was in St. George's West near the airport, and the capitol city of Hamilton was on the other side of the island. Colorful tropical flowers were everywhere, and the sea was greenish due to coral reefs. Foreigners could rent a motorbike but not a car.

We stayed in the dormitory at the Bermuda Marine Laboratory for one month. Because it was already known that *Odontosyllis* emitted light with a luciferin and a luciferase, our objective was to investigate further details of the *Odontosyllis* bioluminescence, such as the nature and properties of its luciferin and the luciferase.

Odontosyllis is a 1–2 centimeter long worm that comes up to the surface of the sea for only a few minutes, one hour after sunset, and only for a few days after each full moon. First a female worm emitting a green light shows up on the surface and makes small circles; then a few seconds later, a number of male worms appear and rush toward the female worm. Their activities look like green fireworks, except that the movement is in the opposite direction, and it happens all over the surface of the sea at the same time.

We collected the *Odontosyllis* using several collectors on a research vessel. It was necessary to plan very carefully because of the short timespan within which the worms would appear. We made deep nets with Dacron gauze and attached them onto hand nets equipped with long handles, so that we could scoop up worms continuously. We had decided in advance how best to handle the collected worms, to minimize the loss of luminescence activity during handling, and we had also obtained 50 kilograms of dry-ice from the United States mainland.

Our collection was successful under favorable weather conditions, obtaining 200 grams of worms in the three days of August 18–20. According to my records, the full moon fell on August 16; on the first collection day on August 18, it went like this: *sun set at 7:02 pm, start collection at 8:00 pm, collection finished at 8:15 pm.* We returned to Princeton at the beginning of September, avoiding the Labor Day congestion, and began to purify our *Odontosyllis* luciferin. We soon learned that, for the bioluminescence reaction of *Odontosyllis*, in addition to luciferin and luciferase, magnesium ions and cyanide were needed. It is puzzling why toxic cyanide is needed for the bioluminescence of *Odontosyllis*. The reason remains unknown, even now.

My U.S. visa was expiring in 1963, and I expected to go back to the position of assistant at the Pharmacy Department at Nagasaki University, from which I had been on leave during all my time in Princeton. However, Professor Hirata offered me a position as an Assistant Professor at the Water Science Institute at Nagoya University, where I would also be able to continue my research in bioluminescence. I consulted with Dr. Goro Kobayashi, the head of the Pharmacy Department at Nagasaki University. Dr. Kobayashi considered that the position at Nagoya University would be advantageous to my future and recommended that I take the position.

CHAPTER 5

A Risky Endeavor, Raw Jellyfish, and Mysteries of Bioluminescence

My wife and I left Princeton in early August of 1963, to return to Japan. Travelling from Philadelphia to Los Angeles by train, we stopped at St. Louis, Missouri, and Denver, Colorado, where we rented a car and climbed Pikes Peak. We met Dr. Fay in Los Angeles. At Long Beach, we boarded *Brazil-maru*, a cargo-passenger ship between Japan and Brazil, bound for Yokohama via Los Angeles and San Francisco. We chose that ship because it carried large amounts of luggage free of charge. When our ship docked in San Francisco, we met my friend Dr. Shizuyuki Yamada, who was studying at Stanford University, and he kindly gave us a sightseeing tour around the city. *Brazil-maru* was a 10,000 ton cargo-passenger boat with only a few passenger rooms, but our room was spacious and we had a very comfortable trip to Yokohama. Every day we took our meals at the same table with the ship's captain, the ship's doctor, and a couple of Dutch scholars who were planning to study old Japanese music.

5.1 Back in Nagoya: 1963–65

I arrived at the Water Science Institute in September 1963 as an Assistant Professor, working under Professor Tadashiro Koyama.

Professor Koyama told me that he planned to retire in a few years, and wanted me to take his position at that time. My future now seemed to be set. He found me and my wife an apartment in the Hoshigaoka area, on the third floor of a five-story building. The apartment had two rooms plus a dining kitchen, and was well built for that time, with a rent close to 20,000 yen per month. I was shocked, however, when I received my first monthly salary. It was only about 30,000 yen, about one eighth of what I had received at Princeton, although I foresaw no immediate problem since I had accumulated considerable savings.

A few months later, Professor Hirata came and asked me to help his graduate student, Yoshito Kishi (later a professor at Harvard), to discover the chemical structure of *Cypridina* luciferin. So, that summer, Kishi and I went to Setoda in the Inland Sea, near Hiroshima, for mass collection of *Cypridina*. In this scenic area, citrus trees grow all over, near the temples. Setoda is one of many islands and citrus fruits are the main produce of these islands.

We took some ten students with us to assist in collection. Our collection tool consisted of an enameled bowl, 20 centimeters in diameter, covered with a round piece of cotton cloth that had a 2-centimeter hole at the center and a rubber band sewn in around the edge of the cloth so that we could easily take it off and put it back onto the bowl. We tied a long cord to strings attached to four spots on the edge of the bowl. To start collecting, we put a small piece of fish or a fish head in the bowl, placed the cloth cover on the bowl, and then hung it by the cord down to the bottom of the sea. After thirty minutes or an hour, we pulled the bowl up, removed its cloth cover, and then poured the sea water containing *Cypridina* over a piece of Dacron gauze to drain off seawater from the *Cypridina*. We used thirty to forty bowls for our mass collection. *Cypridina* is extremely voracious; only bones and skin will remain within if the bowl is left at the bottom for a few hours. Many kinds of collection tools had been made for mass collection of *Cypridina* during the war. The ones we used were made by Mr. Hajime Miyamoto and Mr. Samon Miyamoto of Setoda, in the Hiroshima prefecture.

We froze the collected *Cypridina* immediately with dry ice. We brought back about 5 kilograms of frozen *Cypridina* to Nagoya after collecting them for about two weeks. Then we extracted, purified, and crystallized the luciferin from the frozen *Cypridina,* following the new method developed in 1960. Using that luciferin, Yoshito Kishi successfully determined the structure of *Cypridina* luciferin the following year.

Our first child, our son Tsutomu, was born in October 1964 at Nagoya during the Tokyo Olympics. Around the same time, my wife's father Tomoo Okubo, who was in Nagasaki during the atomic bomb blast, died of cancer. He had had to search around the city for several days looking for my wife and her elder brother Akira Okubo immediately after the atomic bomb blast and saw so many miserable scenes there. He was looking forward to my son's birth very much; he passed away in November, fifteen days after Tsutomu's birth. My father-in law was in the late stages of cancer and in bed at the hospital when he received the news that his grandson was born, and Akemi heard from her mother that he was very pleased. But he never had the chance to see Tsutomu.

5.2 Glow Worms in New Zealand

In February 1965, I traveled to New Zealand to study two bioluminescent organisms, the cave worm *Arachnocampa* and the freshwater limpet *Latia*, with research funds from JSPS (Japan Society for the Promotion of Science). I went together with Dr. Yata Haneda, the director of the Yokosuka City Museum. We arrived in Wellington, New Zealand, via Hong Kong and Sydney, and then rented a car and drove toward Auckland. On the way, we briefly investigated a giant luminous earthworm at Palmerstone North, and saw the glow worm *Arachnocampa* at the famous cave there, although we were not permitted to collect any because the glow worm cave was an important tourist attraction.

In Auckland, we met up with Dr. Johnson who came from Princeton University, and we collected *Latia* in a shallow stream in

Waitakere. *Latia* is a univalve shell about 5 millimeters across, which is found attached onto the surface of rocks. Upon stimulation, the limpet produces a slime that emits yellowish-green light. I found that its luminescence was produced by a luciferin-luciferase reaction and was highly suitable for our research; thus we decided to collect as much *Latia* as possible. According to my records, we collected 2 kilograms of *Latia* in eight days beginning on February 24, 1965, which we froze with dry ice and took back to Nagoya.

The glow worms, which we first saw in Waitomo Cave, are the larvae of the dipteran insect *Arachnocampa*. The larvae are about 2 centimeters long and have a light organ at their tail ends which emits blue light. They stay on a spider-web-like network of slime made on the cave ceiling, and suspend long thin threads of slime, about twenty threads per larva, to "fish" for smaller insects. In Waitomo Cave, the numerous larvae on the ceiling look like stars in the night sky. We collected only a small number of *Arachnocampa* in a cave at Waipu near Auckland. The bioluminescence of *Arachnocampa* requires ATP like the firefly bioluminescence, but it appears unrelated to the firefly luciferin.

In early April, Dr. and Mrs. Johnson arrived in Kobe by ship from New Zealand, intending to stay in Japan until the International Bioluminescence Conference to be held at Hakone in September. Dr. Johnson was a co-organizer of that conference together with Dr. Haneda. I met with Dr. and Mrs. Johnson at the port of Kobe and took them back to Nagoya. Since the University had no accommodations for foreign visitors, I found a small apartment for them. Toward the end of April, I took them to Uozu Aquarium in Toyama prefecture. They saw some live, swimming firefly squid (*Watasenia*) and also the photographs of Newton Harvey, a pioneering scholar from Princeton in the field of bioluminescence studies, taken when he visited Uozu in 1916. Since Mrs. Johnson became bored with the long stay in Nagoya, I invited the couple to an old Japanese-style restaurant at Kakuozan in the eastern section of Nagoya; Kakuozan extends from the Nittai-ji temple, whose name ("Japan-Thai temple") commemorates the donation of a relic of the Buddha, enshrined there, by the Thai king. I also took them fishing with cormorants on the Nagara River, to see the Mikimoto Pearls on the Shima peninsula, and on an overnight trip to Yunoyama hot springs. They were both delighted.

After the international conference ended, I decided to return to Dr. Johnson's lab at Princeton University. The reason for my decision was not my low salary in Japan. Based on two years' experience at Nagoya, I judged that I lacked the ability to accomplish top-level work in both my different fields, water science and bioluminescence. So I had to choose.

By the Glow Worm Cave, New Zealand (1965).

One practical problem for me was how impossible I had found it to clarify the mechanisms of aequorin luminescence, and I knew that if I could not solve those scientific mysteries, aequorin would end up being regarded as merely a peculiar protein that emitted light in the presence of calcium ions. In addition, some scientists at the time still did not even believe that a protein like aequorin could exist. I also thought that if I didn't work in bioluminescence, there would not be much progress in the chemical understanding of bioluminescence in the future. It seems that the spirit of bioluminescence had taken possession of my thoughts ever since my study of *Cypridina* luciferin.

Akemi and Tsutomu with the Johnsons, Yunoyama hot springs (1965).

After I made the decision to return to Princeton, Professor Cormier of the University of Georgia visited me and urged me to come to Georgia instead, but I was sorry that I could not change my plan of going back to Princeton.

Staff of the Water Science Institute, Nagoya University, 1964 (Prof. Koyama: front, third from left; me: front, far right).

Although the members of the Water Research Institute were all reluctant to approve my plan, mainly because they foresaw that I would have a very hard time if I went back to the United States, Professor Koyama and Professor Hirata allowed me to leave Nagoya. Over forty years later, I now think that the attitude of the members of the Water Research Institute toward my choice at that time was absolutely right, and that the decision I made was a chillingly risky one, especially considering that I would soon have held a position as a full professor if I had stayed at Nagoya. Fortunately, I was able to achieve adequately beneficial progress in the field of bioluminescence, so I do not regret my decision. If I had stayed in Nagoya, the puzzle of aequorin bioluminescence would not have been solved, and it is doubtful that the study of GFP would have developed into its present status.

5.3 Back to Princeton: My Family Grows

I resigned from Nagoya University on November 30, 1965. My wife and I, taking our one-year-old son Tsutomu, first flew to Los Angeles to obtain some specimens of the marine parchment worm *Chaetopterus. Chaetopterus* is a grotesque-looking worm made up of feathery segments, which lives in a papery U-shaped tube, around 15 centimeters long, 1 centimeter wide. Upon stimulation, its whole body emits a weak blue light and a section next to the head secretes a brightly luminous slime. The tube is buried in the mud at the bottom of the sea, with only the ends visible. Since its luminescence is not visible outside the tube, it is puzzling why this worm is luminescent.

I asked Dr. Rimmon Fay to obtain *Chaetopterus*, and he willingly collected several thousand specimens for us. Dr. Fay sublet a little beach house for my family for a few weeks at Marina del Rey by Venice, Los Angeles where he had his office. The cottage was surrounded by sand and we could see the surf beyond a quarter mile of the sandy beach. It must have been a paradise in warm weather, but it was not so attractive in December. My son was not old enough yet to enjoy the abundance of sand. I heard that the owner of this cottage was someone connected to Hollywood, but I do not remember who. It was a very quiet and rather lonely place more than fifty years ago. I heard that in later years, as the area around Santa Monica Bay was further developed and the beautiful waters became increasingly clouded, Dr. Fay, who saw those ecological effects firsthand, fought against the marine and shore pollution by industries in the area.

Through arrangements made by my good friend Dr. Ozaki, who was at Caltech at the time, the wives of three Japanese students helped us to dissect the worms, to cut off only the part that secretes slime and freeze it immediately with dry ice. I took this frozen material to Princeton University, where I then began the purification process of the bioluminescent substance of *Chaetopterus*.

In mid-December my family moved into one of the university apartments at Stanworth Drive on Route 206. The apartment was furnished and equipped thoroughly by Professor and Mrs. Johnson. Most of the furnishings were purchased from Johnson's former assistant who had decided to join the Peace Corps. The Stanworth apartments had a playground and provided an excellent environment for families with children. My children enjoyed it until they were grown up nearly into their teens. Although we moved to a larger, three-bedroom apartment there after our daughter Sachi was born in 1969, we did not leave the Stanworth apartments until we departed Princeton many years later.

I requested that the university obtain a permanent visa for my family and me. It was a change from an H–1 visa, and I obtained the permanent visa relatively easily by the end of 1966. I did not have a driver's license yet and depended on my wife's driving. Akemi had practiced her driving a little in Japan and acquired her license when we were first in Princeton; two of her younger brothers were skilled drivers. But once I decided to live permanently in the United States, I realized that I also needed a driver's license. First, Dr. Johnson gave me lessons a couple of times, and then I went to the Motor Vehicles Office to obtain a temporary license after passing a written test. Then, I did my driving practice a few times, with my wife sitting at my right, at a new development area where only a few houses had been built. During my driving test, I could not parallel park smoothly, but I still received my license.

At the time we owned a smart Pontiac, but it had low body clearance and was not very comfortable for a long drive such as a transcontinental trip, so we traded it for a large, powerful Buick LeSabre after four years. We used two generations of the Buick, during which time we made ten transcontinental round trips. At one time, we hit a large deer. When my wife was driving in Montana at 75 miles per hour, a deer suddenly leaped onto the road and the right headlight hooked the head of the deer, causing its body to bang into the front right door. The door was warped and could not be opened. The deer died instantly and fell into the roadside ditch. We wanted to make repairs, so we went to the sheriff's office in the next village to report the accident. The sheriff said that he must go to the spot to kill the deer if it was still alive. I told him that the deer was dead and in the ditch, so he was glad not to have to attend to it. We made temporary repairs so that we could use the right door, and changed the door for a new one after returning to Princeton.

In 1984, we switched to a smaller car, a Chevy Celebrity, because we did not expect to drive to Friday Harbor anymore. We found this to be the worst car we ever owned. The car leaked oil from its head gasket, but the repair service made only a superficial temporary fix, and not a real repair. In addition, the car often went out of control on snow. I had tried to use American cars because I was living here, but I finally gave up on American cars and switched to a Subaru Legacy in 1993, a 4-wheel drive car that could drive on snow in New England. I am unusually tall for a Japanese person, but with this car I would not bump my head on the ceiling. (Whenever I travel to Japan I have to be careful not to bump my head on low ceiling rafters and doorways, especially in older houses like my family property, that were intended for shorter people.)

During these years in Princeton, our young family grew. Tsutomu, as I mentioned, was born in 1964 in Nagoya, in the Year of the

Dragon like myself. His name means "endeavor," and in that trait, too, his character may resemble my own. At stops during our transcontinental trips when he was young, he liked to peek into roadside ditches. When his aunt took him to an aquarium in Nagasaki, he was more interested in studying the pipe system that transported the water instead of looking at the fishes. This surprised our relatives.

For one year before entering primary school, he attended the preschool of the Princeton Institute for Advanced Study, the Crossroads Nursery School, which was started through the leadership of Mrs. Oppenheimer (wife of the famous nuclear physicist) in the late 1940s. By the time my son enrolled, it was located facing Einstein Drive. Albert Einstein had been one of the original members of the Institute for Advanced Study, where he continued his affiliation for more than twenty years until his death in 1955. Therefore, most of the children there were children of scientists who had come to Princeton from all over the world, and the teachers were very used to handling kids from such backgrounds. Sachi had many friends there and enjoyed the program when she attended, but I was not sure that it worked as well for Tsutomu, who from a very young age asked too many questions, most of them beginning, "Why…?"

When I gave him a book on Isaac Newton, I had to explain the contents to him in the evening every day—we talked for many days about gravity, friction, elements, and so on. He could hardly wait for this after-dinner hour of learning, eagerly opening the book to the page following the one we had discussed the previous day even before dinner was done, so that I often had to finish my meal hurriedly. We settled into a regular routine where, after dinner, he asked me to explain pages and pages in the book. Probably this was the beginning of his love of physics.

As a toddler, he liked to listen to a record of Mother Goose songs set to music. When his toy record player broke down, he was very excited to be able to see inside the mechanism of the player. When he was in third grade, we bought for him the twenty-four volumes of the *Encyclopedia Britannica* that he wanted. It became his favorite book.

My wife, when she recently heard that our granddaughter was finding kindergarten too easy, remembered vividly how Tsutomu's teacher in Princeton had said that he often did not join his classmates but was reading an encyclopedia all day long in the corner of the classroom under his desk.

Tsutomu became best friends with some neighbors who lived up the hill from us, Preston and Helen Tuttle. Preston, whom we always thought of as Preston-san, was an ardent naturalist. His house usually contained rescued birds or other creatures, and natural foods that were not so readily available before the natural-food craze (Sachi and Tsutomu first tried yogurt at his house), as well as numerous books on natural history and philosophy. He willingly answered Tsutomu's many questions and encouraged his interests. He was a great mentor and was learned in many subjects. We met some of his students, as well; one of them kindly took Tsutomu and Sachi to the movies a few times. In later years, during his decline in health, after a heart operation and a stroke, my family and I folded and strung 500 origami cranes for him as a wish for good luck and good health. After we left Princeton, Akemi continued to correspond annually with his wife Helen until she passed away. She was a very active and well-read woman, as well as very kind.

When Tsutomu was in fifth grade, *General Chemistry* by Linus Pauling was his favorite book. His interest in computers began when he heard a talk by a scientist from RCA, and he already had a considerable knowledge of computers at thirteen years old when he entered high school. (He explains how he gained his knowledge of computers in Chapter Four of his own book *Takedown,* so I will not repeat the story here.) In the 1970s, almost nobody knew about computers except the experts in the field. Tsutomu used to stay in the computer center of Princeton University after school, and usually did not return home until after midnight; I had to go there to look for him several times. Graduate students and postdoctoral fellows used to ask him for help, and they often called our home asking him to assist them even in the middle of the night. He helped and was paid by several professors during this time.

While he was very engrossed in learning about computers, at high school, on the other hand, he did not like to be taught material that he already knew, so he skipped attending classes like chemistry and English. He also missed his English class frequently because he would oversleep the next day after staying at the computer center until the early hours of the morning. So Akemi and I were called to his school several times. We had some family tensions and arguments about his actions, because I wanted him to complete high school although I sympathized with his desire to learn as much as he could at Princeton's supercomputing center. In the end, he did not graduate from high school due to a record of insufficient attendance. As a result, he could not enter MIT or Carnegie Mellon as he had wanted, but happily he was accepted to the University of Chicago and Caltech the following year.

Our daughter Sachi, more than four years younger, was born much more easily than Tsutomu, whose birth was long and exhausting for Akemi—it took longer than a day. It was well after midnight, possibly near dawn by then, when I drove her to the hospital on the next street on a chilly winter night. I had to return home immediately because we had left Tsutomu there. Sachi was born early that morning. The day was very cold and bright. As I recall, the doctor said that Akemi could go home the following day, but she asked to stay an extra night to rest. When she came home, she thought I looked rather tired from taking care of Tsutomu for those two days. I had chosen to keep him with me rather than take him to nursery school.

Sachi was quite different from Tsutomu. She studied well what she was taught at school. For a few years, she was in the same classroom as a son of the Governor of New Jersey, who lived just several blocks away in the old governor's mansion. They did not share the same group of friends, but she was delighted at the idea of going trick-or-treating there. One year she dressed as a jellyfish for Halloween, in a costume made of old translucent white curtains over a cardboard frame that I crafted for her. Every Christmas Tsutomu and Sachi particularly looked forward to the gifts that Dr. Johnson and his wife would give us—the Johnsons delighted in finding a splendid assortment of toys, gadgets, and books for my children, especially when they were younger. Akemi learned to bake American foods like buttery croissants and cream puffs using recipes she got from Mrs. Johnson, who would give us home-baked treats every Christmas; Sachi and Tsutomu each had their favorites. Sachi made friends with a number of children of Princeton faculty and visiting scholars over the years. One friend, whom she was very sad to see move away after first or second grade, she happily ran into again much later in graduate school. She was accepted by several universities, but after consideration, in the fall of 1986, she chose Stanford University; she was 17 years old at that time because she had skipped eighth grade.

We took our children back to Japan three times in their childhood so that they would know their relatives. During those years, we always talked about returning to Japan eventually. The first time we made the trip, Sachi was still a baby, barely toddling around and very unhappy to be stuck on airplanes for so many hours. We insisted that they learn Japanese, and we spent two months in Japan before Sachi started kindergarten. During that stay, I worked on luminescence research on various organisms, particularly *Watasenia* (firefly squid), with the Japanese scientists Dr. Toshio Goto and Dr. Shoji Inoue at Nagoya University. Tsutomu attended a Japanese school. I think he was not so happy there, since he, like me, has always been a bit impatient with following strict rules. He was not particularly good at conforming to expectations in school in Princeton, either.

In Japan, we lived in a small house with the traditional toilet facilities (a deep pit in the floor with a wooden lid to cover it) and slept on futons on the *tatami* (woven straw mat) floor. During the day we would usually walk up the road to Akemi's mother's house, where Sachi would play with my three-year-old niece Megumi and baby nephew Makoto. She had to learn not to poke her fingers accidentally through the *shoji* screen walls partitioning the rooms; small holes were often repaired with cherry-flower shaped patches of rice paper glued on with rice paste. Along with several cousins, there were lots of cats and aunts and uncles nearby. Both my children enjoyed the lively family group there. One of Akemi's brothers owned a hardware store, and another loved electronics, and Tsutomu especially loved to go hang around with them. However, their most vivid memory of Japan may have been the new amusement park that they visited while there.

Ten years later, we had moved away from Princeton where Tsutomu and Sachi could go to Japanese school on the weekend, to Cape Cod, and they had started to forget much of the language. They enjoyed talking to their aunts, uncles, and cousins whenever possible, but we all knew that there was little likelihood we would ever move back to Japan now.

By our next trip to Japan, they were old enough to appreciate seeing the atomic bomb museum in Nagasaki, where a display showed the watch that had belonged to their uncle (Akemi's eldest brother), still indicating the time of the bombing, and discovered along with his shoes months after his death from the atomic bomb. My children had grown up hearing stories about the war, and especially about their grandfather's long search to find out what had happened to his eldest son.

Left: My family by the Stanworth apartments, Princeton (1969); Right: by Hells Canyon of the Snake River, Idaho (en route to Friday Harbor, 1971).

5.4 Luminescence Mechanism of Aequorin: Friday Harbor

From Princeton, we continued traveling to Friday Harbor nearly every summer. The first time we skipped a summer, we suffered constantly from the heat and the humidity of summer in New Jersey. My children were also sad not to see their friends that year; they were especially delighted to meet up with my old friend Ozaki's children many summers at the Friday Harbor Laboratories. There, we would often live in cabins or small apartments near the "village green" where the children would race around, feed the rabbits, or play group outdoor games. Once in a while one of the older children would break a window with a ball or other projectile, and then the group would be broken up or supervised more closely for a while. But morning and evening, before they could go and play, Tsutomu and Sachi (when she was old enough) would help me and Akemi collect and process jellyfish for my aequorin studies.

Our jellyfish collectors, Friday Harbor town dock, 1974; from left: Akemi, Dr. Chang, me, Mrs. Chang, Mrs. and Dr. Johnson, Debby, Tsutomu, Sachi.

Why did I decide to invest my time in studying aequorin? The explanation involves understanding the general principles of bioluminescence. While people have been intrigued by living creatures that give off light—bioluminescence—since ancient times, much remains to be understood about the functions, chemical processes, and evolutionary paths involved. Sometimes, the chemical processes give hints as to why and how the ability to emit light developed, and may also suggest useful applications for scientific research or medical study.

The overall process of bioluminescence involves the release of energy, as a bright burst of light, in a chemical reaction such as the oxidation of a molecule. The light-releasing molecule is termed a "luciferin," named after the Latin for "light-bearing" as in the name of the fallen angel Lucifer. The term "lucifer" had become a common noun for a friction match prior to the coining of the chemical term late in the nineteenth century. To early scientists, the bioluminescence may have resembled the lighting of a match or candle. These luciferins can differ among different organisms, but some unrelated species utilize the same luciferin. To understand the chemical process of bioluminescence requires knowing the structure of the molecule that is broken down or oxidized to give forth light, how it is acquired (such as by eating smaller sea creatures that produce the luciferin) and where it is stored, the substance (enzyme) that causes that light-emitting reaction to occur or continue, and the factors that trigger the process. The trigger can be an environmental change or some ion released by the bioluminescent creature to control the light given forth. Many ocean species display luminescence; the reason is often not known. In addition, the color of their light may be affected by a fluorescent protein. All these aspects of bioluminescence can yield interesting insights into nature, and possibly useful applications, if they do not remain utterly mysterious.

Calcium ions had been thought to play important roles in living bodies since the 1950s, but there had been no way to demonstrate that experimentally. However, E. B. Ridgway and C. C. Ashley of the University of Oregon experimentally proved the involvement of calcium ions in the contraction of muscles for the first time in 1967, using aequorin.

Because the importance of aequorin became clear in the fields of biology and physiology, and it was expected to have a wide range of uses, I wanted to clarify the mechanism of aequorin luminescence to facilitate its use. I already knew that the aequorin luminescence was the result of an intramolecular reaction within protein molecules; therefore, I expected formidable difficulties in my study, but I decided to persevere. It ended up being an unexpectedly large undertaking, requiring about twelve years to obtain a structural model of aequorin, as I explain below.

5.5 How Do You Prepare Jellyfish?

Extracting the chromophore from aequorin was essential to determining its luminescence mechanism. The chromophore is the portion of a molecule that produces its color. Chromophores absorb certain wavelengths of light and reflect others, producing different colors depending on which wavelengths are reflected. Bioluminescence compounds can have light-emitting chromophores, which both produce light and absorb certain wavelengths; they can also have fluorescent chromophores, which absorb wavelengths of light produced by other sources. Extracting the light-emitting chromophore from aequorin proved to be impossible, however, because any treatments we tried on aequorin always triggered a luminescence reaction (which would break down the chromophore). Therefore, we gave up on trying to extract the light-emitting chromophore, and tried to extract a fluorescent chromophore instead, because a fluorescent chromophore is essential for the emission of light. Fortunately, we found that the denaturation of aequorin in the presence of mercaptoethanol produced a blue fluorescent substance. Since this substance had an absorption maximum at 350 nm, we named it AF350 in 1969, and decided to determine its chemical structure. Because AF350 was only a very small part of the aequorin molecule, we would need 100–200 mg of aequorin for a single chemical experiment that would allow us to study its structure. Thus, we had to collect about 50,000 *Aequorea* and extract aequorin at Friday Harbor each summer for the next ten years and more.

Except for days of bad weather, when the jellyfish would withdraw deep into the ocean, we collected at least 2,000–3,000 jellyfish every day, cut off their rings of luminous tissue, and extracted aequorin from the rings. Since a bucket would usually hold around a hundred jellyfish in seawater, our goal was thirty or more full buckets each collection day. We took the buckets back to the lab in cars or motorboats, and transferred the jellyfish to temporary holding tanks with circulating seawater. Then we would hose off and reuse the buckets. Often, as soon as we arrived on the island for the summer, we would buy buckets at the hardware store to replace the ones that had worn out or been lost. We liked to draw jellyfish pictures on them to help identify them as our property.

At Friday Harbor, we used to start jellyfish collection at 6 a.m., have a quick breakfast, and then start the ring cutting. We extracted aequorin from the rings from 1 p.m. to 5 p.m., and again collected jellyfish for the next day from 7 p.m. to 9 p.m. However, we were soon confronted with a problem; we could not cut the rings fast enough. Cutting a ring with a pair of scissors took at least one minute, so that it was not feasible to cut 3,000 rings even with several helpers.

This problem was solved by Dr. Johnson's ingenuity. He found two used ten-inch meat-cutting blades at a hardware store (at this time, the main stores I remember at Friday Harbor were a hardware store, a drugstore and a food market, as the town was still very small), and built jellyfish cutting machines: He replaced the meat-loading part of the meat-cutter with a horizontal black Plexiglas board, and an electric motor slowly rotated the blade. After placing a jellyfish on the Plexiglas board, we spread the edge of the jellyfish (it was about the size and shape of a plump sunny-side-up egg) and pushed it against the rotating blade while rotating the jellyfish; a thin strip of its margin would thus be cut off. To rotate the jellyfish, we used a small disk-shaped "fork" with spikes. (At first it was a simple wooden disk with nails. The final version was a sleek, rustproof Plexiglass and stainless steel instrument custom made by a skilled machinist recommended by Dr. Shinya Inoue and suggested by my daughter who was at that time his summer lab assistant at MBL; I think it cost $300 for a pair.) After various improvements, we could cut 1,200 rings per hour using two of these cutting machines.

The Johnsons demonstrating the jellyfish cutting machines.

In the years that followed, Akemi and several long-term assistants, Debbie, Liz, and Laura, who were also great favorites of our kids, became especially skilled cutters. Tsutomu and Sachi liked to watch them work, until they were old enough to help out. We often processed 6,000 jellyfish (300 kg) in one day, and our lab looked like a jellyfish factory. It was filled with the fumes of jellyfish, which also clung to our hands. To avoid disturbing other researchers, we moved our lab to a nearby storage building known as the Gear Locker.

In later years, we did much of our jellyfish collection at the Friday Harbor town dock, the Marina, which grew tremendously over the years that we were collecting. Every year my children were excited to pass under the "Port of Friday Harbor" sign on the first day of collection, and curious to see what new branches had been added to the dock. My son, in particular, liked to go off alone to collect buckets full of jellyfish at the far edges of what was then the "B" dock, and later expanded through other letters of the alphabet. When he was younger, my wife would go with him, while my small daughter would stay close by one of us. By the time she was five years old, I made her a special, extra-short jellyfish collection net with a bamboo handle, so that it would float if she dropped it. Sometimes kind people on the boats would help her get it back if it fell near their boat. My son also had a long, special bamboo net of his own. Dr. Johnson equipped the rest of the collectors with matching long-handled nets with wooden poles painted orange and black (for the Princeton Tigers) attached to heavy-gauge steel wire; these would often sink if dropped accidentally into the ocean. These collection nets are now museum pieces. Just recently, in response to a request for a permanent display, we mailed the last one to Japan, to the National Museum of Nature and Science in Tokyo. UPS charged more than five hundred dollars for shipping.

On windy days at the dock, Akemi had to pin her hat on carefully, or else it would blow away, and someone would need to take a motorboat or rowboat to fetch it before it sank. The ferry terminal was close by, and large waves would rock the town dock when large ferries passed by, so that we would have to be careful not to slip. But sometimes those same waves would wash jellyfish up to the surface. My children loved to predict the size of the waves and identify each ferry by name: *Vashon*, *Nisqually*, *Klickitat*, *Kaleetan*.

As we collected *Aequorea*, residents of the docked boats would repeatedly ask, "Are they edible?" "How do you cook them?" "What are they for?" Usually we replied, "We use them for research." Some of my more mischievous teenage assistants, tired of answering the same questions twenty times a day, would at times pretend that they would become the base for grape jelly, or else other more exotic substances. A story that I have sometimes told is about one old lady, poking her head out from one of the boats to ask, "How do you cook them?" When I replied, "We don't cook these," she exclaimed with a distasteful expression, "Do you eat them raw?" Then she disappeared back into her boat without giving me time to explain.

Owing to improvements in the purification techniques, the amount of purified aequorin we produced increased every year, and we learned more detailed properties of aequorin. In 1972, following several years of purification and research, we finally determined the structure of the part of the aequorin molecule that we had labeled AF350. Its structure, however, surprised me greatly. It contained the same partial structure as in the *Cypridina* oxyluciferin that was formed from *Cypridina* luciferin in its bioluminescence reaction. Because of this unexpected similarity, I thought that the luminescence reaction of aequorin might resemble that of *Cypridina* luciferin. With this in mind, I examined the product of the calcium-triggered bioluminescence reaction of aequorin, and found a blue fluorescent substance. This blue fluorescent substance was found to be acylated AF350, that is, bonded with an acyl (a type of molecular group that often occurs naturally in cellular processes). Thus this blue fluorescent substance more closely resembled *Cypridina* oxyluciferin than AF350.

In 1974, I ultimately concluded that aequorin bioluminescence and *Cypridina* bioluminescence employed the same type of chemical reaction, and drew the structure of the light-emitting group of aequorin based on the structure of *Cypridina* luciferin. In 1975, I named the light-emitting group of aequorin "coelenterazine" and the acylated AF350 "coelenteramide." I would like to emphasize here that the structure of coelenterazine was conceived because the chemistry of the *Cypridina* luminescence had been known already.

Coelenterazine was isolated in 1975 as the actual substance from the firefly squid *Watasenia* by Professor Shoji Inoue, at Meijo University, and then synthesized. Using the coelenterazine synthesized by Professor Inoue, I was able to prove that the light-emitting substance in aequorin was coelenterazine, by regenerating the active form of aequorin from the spent aequorin that was formed by the calcium-triggered luminescence.

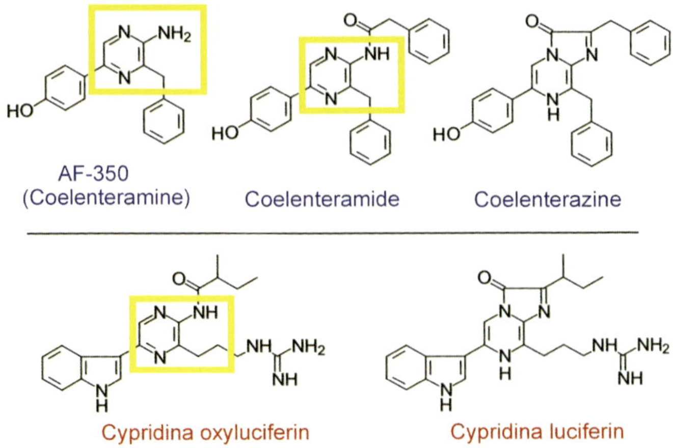

Structure of AF-350 (part of aequorin molecule); coelenteramide (AF-350 bonded to an acyl), produced when aequorin emits light; coelenterazine (the light-emitting substance in aequorin). Similarity of coelenteramide to *Cypridina* oxyluciferin (produced when *Cypridina* luciferin emits light). Below right: taken when starting jellyfish collection at 6 am, when still dark.

Similar to how the converse functions of multiplication and division can allow one to check a simple equation by running it in reverse, chemists can "check" a proposed chemical reaction by showing how it can be run backwards to regenerate the original substance altered in the reaction.

There was one more problem remaining to be solved in the luminescence reaction of aequorin. The problem was to understand why the luminescence of aequorin did not require oxygen despite the luminescence of coelenterazine requiring oxygen. To answer this problem, I performed various chemical experiments, and then concluded in 1978 that aequorin contains the peroxide of coelenterazine. Namely, aequorin does not require added oxygen to emit light because it contains oxygen in the molecule. Thus, I had accomplished my first objective.

5.6 My Research on Subjects Other Than Aequorin: 1965–1978

Although aequorin presented my most consuming challenge, I continued to work in other directions during that period. Perhaps in part because of my untraditional educational path, I had learned early on the value of piecing together knowledge from different directions to shed light on biochemical processes that were not yet thoroughly understood. Between my return to Princeton in 1965 and achieving a general understanding of the aequorin luminescence reaction in 1978, I thus undertook to research the bioluminescence of various luminous organisms, including the following five kinds, often choosing my direction of study based on material at hand that had been obtained during my previous trips.

5.6.1 Luminous limpet *Latia*

Starting work in 1966, I elucidated the structure of the luciferin of the New Zealand fresh water limpet *Latia*, and investigated the properties of *Latia* luciferase. The luciferin turned out to be a sesquiterpene (a compound based on carbon and hydrogen that often gives forth a strong aroma), and would slowly decompose in water. The luciferin emitted yellowish green light in the presence of the luciferase. Thus the light-emitting group appeared to be a flavin—a type of compound named after the Latin word for "yellow"—as in the case of luminous bacteria. However, this has not yet been firmly proved.

5.6.2 Krill

I visited Kristinebergs Zoological Station in Sweden together with Dr. Johnson in August 1966 to study the bioluminescence of the krill *Meganyctiphanes*. Krill are tiny shrimp, 2–3 cm long, with ten small light organs that emit very intense blue light. They live deep in the sea in the daytime, but come up near the surface at night. We stayed at the laboratory at the Station for one month, but we obtained only 150 specimens, and thus could not do a detailed study. Instead, we enjoyed the beautiful seashore and sightseeing of historic Viking remains. On the way back, we stopped at the Millport Marine Station in Scotland, and obtained 10 kilograms of *Meganyctiphanes*, which we took back to Princeton on dry ice.

At Princeton, we isolated two components from *Meganyctiphanes* that were essential for light emission, a colorless compound that exhibited a strong blue fluorescence, which we labeled "F," and a protein, "P." Both F and P were highly unstable, and had to be extracted and purified at low temperatures as quickly as possible. I spent almost two years in devising a method to fully purify F in less than twelve hours. First we mistakenly thought that P was the luciferin and F corresponded to the luciferase, but later we studied another kind of krill, *Euphausia*, and found that F was the luciferin and P was the luciferase. F contained four pyrroles in its structure resembling chlorophyll.

The total structure was determined later in 1988, by the late Dr. Hideshi Nakamura who was working at Professor Kishi's lab at Harvard at the time. Dr. Nakamura carried out the extraction and purification of F at my lab, at that time in Woods Hole, and I was strongly impressed that he mastered the delicate and complex procedure very quickly, almost instantly. It was a great grief to me that he died in an accident while diving in Okinawa to collect specimens in 2000.

I had awaited further breakthroughs through his research and hoped that he would carry on my work after I retired; he picked up some specimens from my lab just one month before his accidental death.

5.6.3 *Chaetopterus*

We extracted a bioluminescent substance from the material we obtained in 1965 in Los Angeles. But the extract was slimy and viscous, and difficult to purify, probably because it contained a high concentration of nucleic acids. Upon finally completing purification and leaving the purified material standing for a while, however, we found that it gradually crystallized by itself. We studied its properties and learned that it was a bioluminescent protein like aequorin, that is, neither a luciferin nor a luciferase. We still lack full information concerning the nature of its light-emitting chromophore.

One funny incident occurred when we went back to collect more *Chaetopterus* years later. By then my children were both old enough to help cut open the tubes to access the worms. But sometimes the worms would stretch thin enough that we could not tell for sure where the worm would be within its opaque tube, and what would slither out with the first incision, so after a few unpleasant surprises, my daughter Sachi was on edge and very unhappy with her task. We kept the mass of tubes in a large tank filled with seawater on the table before us as we worked. Suddenly, as I reached in to extricate another tube from the tangle, there was a large splash and a disturbance: a small black creature jumped high into the air, startling all of us and especially my already nervous daughter. Then we all laughed. We had frightened a tiny octopus, very cute, which we rescued. Sachi was much more charmed with the octopus than with the worms; to this day, she hates *Chaetopterus*. When working later as an assistant in the summer at Dr. Shinya Inoue's lab at MBL, she was happy to feed the sea urchins and keep their tanks clean, but was not at all enthusiastic on those occasions when she needed to fetch *Chaetopterus* to his lab from the Marine Resources building.

5.6.4 Dioxetane hypothesis

Although it had been believed that the luminescence reaction of firefly luciferin involved a dioxetane intermediate, Dr. Marlene DeLuca presented a new hypothesis in 1970 that did not involve a dioxetane, based on experiments with an O–18 isotope. I knew by intuition that there must be an experimental error in DeLuca's experiments. However, she was an established scientist, and was the wife of William McElroy, former Director of the NSF and, at the time, the Chancellor of the University of California San Diego, who held a strongly influential position in the field. I met them in 1972 at the International Chemiluminescence Conference held at the University of Georgia. They knew me as the discoverer of aequorin but showed no interest in other subjects, so far as I can recall.

I wished that Dr. DeLuca would realize her mistake and submit a correction paper. Thus, I published four papers between 1971 and 1974 on the bioluminescence reaction of *Cypridina* luciferin that implied her error. My efforts were ineffective. Moreover, three follow-up papers involving four laboratories were published supporting DeLuca's hypothesis. In the end, I decided to repeat DeLuca's experiment using the firefly luciferin, instead of *Cypridina* luciferin. To prepare the luciferase, we collected American fireflies near Carnegie Lake of Princeton University, and Dr. Toshio Goto sent us Japanese fireflies. The results of our experiments clearly proved the dioxetane hypothesis, not the Deluca hypothesis. The error in DeLuca's experiments had been caused by contamination with the carbon dioxide that exists in air. Thus, the problem was solved, but my paper submitted to *BBRC* was never accepted during the period when McElroy was an editor of that journal. Since the relationship between Dr. McElroy and Dr. Johnson, both Harvey's students, appeared not to be especially good, that might have had some effect on the situation. Or perhaps the journal editors thought my experiments were flawed in the face of the four papers that had supported contrary results; I don't know. I prefer to deal with pure science, and have often wished that there were no politics involved in scientific study. It is a fact of science that progress often involves errors (or favored hypotheses based upon incomplete evidence) and their eventual correction, as I had learned even in my original study of aequorin. Even Einstein made mistakes; careful correction seems to me one of the cornerstones of basic scientific progress.

5.6.5 The firefly squid *Watasenia*

I went to Uozu, Japan with Dr. Inoue, of Meijo University, in 1970, to study the bioluminescence of *Watasenia*; I returned to Uozu in 1974. Dr. Inoue was a good friend of mine and also of my dear friend Dr. Goto, so I was happy to have this chance to collaborate with him. *Watasenia* is a small deep-sea squid, about 5 centimeters long. The squid comes to the shallows by the shore in April and May to lay its eggs. Its scientific genus name *Watasenia* was derived from the name of Shosaburo Watase who first described this squid in detail. The squid has numerous minute light-emitting photophores plus large light organs at the tips of two arms that emit strong bright light. We ground the arm light organs with a small amount of water in a mortar. The ground mixture was only dimly luminescent, but when ATP was added, its luminescence intensified markedly. However, when the ground mixture was filtered, and then ATP was added to the filtrate, little light was emitted. I concluded that the direct involvement of ATP in the bioluminescence reaction of *Watasenia* was doubtful. As already mentioned, Dr. Inoue isolated coelenterazine from *Watasenia* in 1975, and he then clearly showed that the luciferin of the *Watasenia* luminescence was the disulfate of coelenterazine, in 1976.

However, Dr. Fred Tsuji, a scientist at the Scripps Institute of Oceanography, reported in 1985 that the bioluminescence of *Watasenia* was caused by the luciferin, ATP, magnesium ions and an insoluble substance, based on studies of a ground mixture of arm light organs. His conclusion was acceptable because he did not specify or discuss the role of ATP. But in 2002, he presented a proposed reaction scheme of the *Watasenia* luminescence reaction that directly involved ATP as a reactant, without substantial foundation. The reaction scheme resembled the reaction scheme of the firefly bioluminescence, involving the decomposition of ATP into AMP and pyrophosphoric acid. I thought that the scheme must be incorrect, and I was concerned that the incorrect scheme would be cited as a possible scheme of the *Watasenia* bioluminescence far into the future, if no one interfered. Fortunately, however, Dr. Teranishi of Mie University demonstrated after lengthy efforts that no AMP or ADP was formed in the *Watasenia* luminescence reaction, showing that the scheme put forward by Tsuji was incorrect.

An important question remained: What was the role of ATP, if it was not directly involved in the light-emitting reaction? Was its role to activate a luciferase? The answers to these questions, which might shed light on other, comparably mysterious chemical systems in nature, are still unknown. However, the crucial first steps, that is, determining what is really known or not known, have been taken, so I am hopeful that further research into the role of ATP in bioluminescence may yield interesting and useful results in science, someday.

5.7 Photoprotein

As described earlier, in 1961, a peculiar bioluminescent protein was discovered from the jellyfish *Aequorea*, and it was named aequorin (Shimomura *et al.*, 1962). This protein emitted light in aqueous solutions by merely adding calcium ions, regardless of the presence or absence of oxygen. The light was emitted by an intramolecular reaction of the protein without oxidation of the molecule, and the amount of light emitted was proportional to the amount of the protein luminesced. These properties do not conform with the definition of luciferin. We first thought aequorin could be an extraordinary exception occurring in nature. However, in 1966 we discovered a bioluminescent protein in the tubeworm *Chaetopterus* that emits light when a peroxide and Fe^{2+} are added; again, the total light emission was proportional to the amount of the protein used. Given the likelihood of many similar bioluminescent proteins still waiting to be discovered in luminous organisms, we proposed a new term, "photoprotein," to designate these unusual bioluminescent proteins (Shimomura and Johnson, 1966).

Thus, "photoprotein" is the general term for bioluminescent proteins within the light organs of luminous organisms that emit light proportionally to the amount of the protein (Shimomura, 1984). A photoprotein could be an enzyme–substrate complex that is more stable than its dissociated individual components, enzyme and substrate. Because of its higher stability, it occurs as the primary light-emitting component in the light organs. For example, the light organs of the jellyfish *Aequorea* contain aequorin, which is highly stable in the absence of Ca^{2+}, while its components coelenterazine and apoaequorin, both unstable, are barely detectable in the jellyfish. (When the aequorin molecule binds with calcium ions, it breaks down into coelenterazine and another substance, named apoaequorin, when producing the jellyfish's luminescence.)

Several different types of photoproteins are now known: for example, the Ca^{2+}-sensitive types from various coelenterates that are popularly termed jellyfish (aequorin, obelin, mnemiopsin) and from protozoa (thalassicolin); the peroxide-activation types from the scale worm (polynoidin) and the clam *Pholas* (pholasin); and the ATP-activation type from a Sequoia millipede, discussed in more detail in the next chapter.

CHAPTER 6

From Mushrooms in Woods Hole to Red Jellyfish in Bergen

For many years, I had worked in the laboratory of Dr. Frank H. Johnson, a suite of two larger lab rooms connected by a small linking chamber where we housed the Bunsen burner. Once our children were both attending elementary school, Akemi joined me in the lab, and as they grew older, Akemi gradually increased the hours she worked with me as a research assistant, leaving the lab in the early afternoon in time to get home before they had walked back from school.

As young children, Tsutomu and Sachi loved to see the lab. I would take them along when I went to check on the lab on Sunday mornings. Usually, when the weather was nice, we would go walk in the nearby garden with a pond, or visit the large water fountain that was one of Tsutomu's favorite destinations as a child. During the rest of the week, they often accompanied Akemi when she drove out to pick me up at the end of the day. Frequently I would give them a "gift"; their favorite was a small plastic bag of tiny glass spheres that we normally used for filtration. They loved to see the sparkling glass and the motion of the bead-sized spheres. (They would take the bag home, and often, on the next day, Akemi or I would take the bag back to the lab, and it would again be the "gift" the next time they came to see the lab.)

Sometimes they would get another special treat—chocolate ice cream served in a small beaker. They each had a specific size of beaker that they would recognize in the glassware case. (Sachi may have used a polypropylene beaker while she was young enough to drop the glass ones.) Sometimes Dr. Johnson would have some surprise ready for them, too. They also loved to see us blow balls on the ends of glass tubes, using the large Bunsen burner in the connecting room. Once, after much practice, I blew and shaped a pair of glass swans, maybe for Akemi's birthday. They joined our display of ornaments on top of the Steinway piano.

6.1 Dr. Johnson's Retirement: Our Move to MBL (1977–81)

In 1977, I moved out of the chemistry building and to the Forrestal complex, where I could continue my laboratory work in an outlying building belonging to the university, located on Route 1 a few miles away from the main Princeton campus. There, I occupied one whole floor at the basement level, so that it was a very spacious lab. Some years earlier, nearby labs had been used for classified physics research on nuclear fusion; Forrestal was named after a former Secretary of Defense.

After Dr. Johnson retired in 1977, I had planned to move to a marine laboratory. First I tried Hopkins Marine Station, Stanford University, where the director, Dr. Pittendorf, agreed to accept me, but he failed to obtain approval from the University. In 1981, however, I moved to the Marine Biological Laboratory (MBL) in Woods Hole, Massachusetts, as a Senior Scientist through the support and kind help of Dr. J. Woodland Hastings, Professor of Biology and Biochemistry at Harvard University, and Dr. Benjamin Kaminer, Chairman of the Physiology Department of Boston University School of Medicine. Dr. Hastings had been a long-time friend of mine, and he and Dr. Kaminer had been friends well before I met Dr. Kaminer for the first time in Woods Hole. Dr. Shinya Inoue also helped welcome me to MBL, and was a wonderful friend and colleague there for many years thereafter. When I moved, I also accepted the position of Adjunct Professor at the Medical School of Boston University.

My research during that period was to have long-lasting impact on both my own life and the long-range application of fluorescent proteins, although I did not expect either outcome at the time. The following is a summary of my work in that period:

6.1.1 The chromophore of GFP

In 1962, we discovered the green fluorescent protein GFP, and reported the properties of GFP in 1974. In 1979, we were finally able to study the chromophore of GFP. The main reason for the slow progress was the very small yield of GFP; the content of GFP in the jellyfish was much lower than that of aequorin. We could obtain only 20–30 milligrams of GFP a year out of 40,000–50,000 jellyfish, each still collected by hand-held nets, whereas we needed 100 milligrams of GFP for one experiment to study its chromophore. So we returned to Friday Harbor Laboratories summer after summer. After collecting the necessary amount, we had to isolate and purify the peptide that contained the chromophore. When we were done, I realized that the absorption spectrum of the peptide was nearly the same as for a compound that I had synthesized while studying *Cypridina* luciferin twenty years earlier. Based on this resemblance, I could identify the chromophore structure of GFP in only six months, much faster than the same study was accomplished for the aequorin chromophore.

The result of the experiment done with accumulated GFP was astonishing. All known visibly fluorescent proteins were a complex of a protein and a fluorescent substance that contains a chromophore, but the chromophore of GFP was found in the peptide link of the protein. Namely, in the case of GFP, the protein itself is fluorescent. It is made of the condensation reaction of the three amino acids: serine, tyrosine, and glycine—that is, these amino acids release water as they bond together. The chromophore structure was confirmed fourteen years later, in 1993, by Dr. William Ward's lab at Rutgers University. The structure makes GFP relatively stable and also appropriate for cloning. This unusual chromophore set the foundation for the future importance of GFP.

6.1.2 Dinoflagellate luciferin

When Dr. Hastings invited me to give a seminar at his lab in 1979, he showed me the preparation of dinoflagellate luciferin which was being purified at his lab. It was blue and looked almost identical to an oxidized sample of krill luciferin. I suggested to Dr. Hastings that his material might have been largely oxidized, and also that it might be worthwhile to compare dinoflagellate luciferin with krill luciferin. He agreed to do the comparison, and sent his student Jay Dunlap, now a professor at Dartmouth University, to my lab to purify dinoflagellate luciferin using the purification method of krill luciferin. The purified dinoflagellate luciferin was practically colorless and exhibited a strong blue fluorescence just like krill luciferin. Although dinoflagellates and krill are biologically unrelated organisms, these luciferins were found to be chemically very closely similar. In 1989, the structure of dinoflagellate luciferin was determined by Dr. Nakamura, who had also determined the structure of krill luciferin a year before. This type of comparative research can thus reveal odd correspondences in the chemical processes produced by nature in ways that do not always fit our current views based around biological divisions.

6.1.3 Luminous scale worms

In 1980, Dr. Marie-Therese Nicolas of France extracted and purified a bioluminescent substance from scale worms at my lab and obtained a protein that emitted light in the presence of superoxide anions. Scale worms are about 2 centimeters long, with the whole body covered by scales. They are clever worms, well adapted for survival. When attacked by other animals, a scale worm can detach two or three scales from its own body and escape to other places, leaving behind the luminous scales as a target for the predator. Scale worms can regenerate their lost scales. *Cypridina* shows similar behavior when attacked by an enemy; *Cypridina* secretes luciferin and luciferase at the same time into seawater, creating a blue-luminescent cloud, while the animal escapes into the darkness of the surrounding area.

6.2 At the Marine Biological Laboratory in Woods Hole

I moved my laboratory from Princeton to the Marine Biological Laboratory (MBL), Woods Hole, in December 1981. With permission from Princeton University, I transported all the equipment and chemicals of my laboratory to Woods Hole on a large trailer truck. The Marine Biological Laboratory is the oldest marine laboratory in the US, and was visited by the emperor of Japan, Hirohito, in 1975. One of the roles of this laboratory is to support summer research for biologists (usually for the three summer months). Thus, the number of staff increases to nearly 1,000 in summer from about 200 in winter. Since its establishment, many Japanese scientists have visited MBL. These include historic names such as Shosaburo Watase, Umeko Tsuda, Hideyo Noguchi, Sakyo Kanda, and Katsuma Dan. So it seemed like an auspicious move for me.

The administration building of MBL is a three-story stone building called Candle House, because it was originally a factory making candles from whale oil. Although the village of Woods Hole is tiny, there are two other large institutions located there, the Woods Hole Oceanographic Institution and the National Marine Fisheries Service of the Commerce Department, along with a branch of the U.S. Geological Survey. Thus this small village forms a very concentrated and dynamic scientific community during the summers. For non-scientists, Woods Hole is known for its ferry to Martha's Vineyard.

MBL has an excellent library, with a five-story book room. In the library, a famous poster written by Katsuma Dan, "The last one to go," is displayed. When the war ended in 1945, the Misaki Marine Laboratory of Tokyo University, which embraced a mini-submarine base of the Japanese Navy, was taken over by occupation forces. At that time, Dr. Katsuma Dan, a lecturer of Tokyo University (later President of Tokyo Metropolitan University) left an appeal to occupation forces written on a sheet of paper with a brush and India ink on the entrance door.

The appeal read, "This is a marine biological station with her history of over sixty years. If you are from the Eastern Coast, some of you might know Woods Hole or Mt. Desert or Tortagus, if you are from the West Coast, you may know Pacific Grove or Puget Sound Biological Station. This place is a place like one of these. Take care of this place and protect the possibility for the continuation of our peaceful research. You can destroy the weapons and the war instruments. But save the civil equipments for Japanese students. When you are through with your job here notify to the University and let us come back to our scientific home." It was signed, "The last one to go." This appeal impressed Colonel Parks of the occupation force, who took off the sheet and sent it back to the United States. It was eventually published in various newspapers and magazines. A scientist who read this article happened to be Katsuma Dan's friend and he thought this paper must be written by Katsuma because he made English mistakes in his statement, and these same sorts of mistakes were often made by Katsuma. Therefore, this paper was sent to Woods Hole and came to MBL.

When I moved to Woods Hole, I transferred the remaining funds of my NSF grant from Princeton University to MBL, which resulted in a significant increase in the funds I could use, even after the payment of the moving expenses. An institution charges a certain percentage of grant money (called an overhead and indirect costs) to manage a grant. The increase in the usable funds was caused by the fact that the overhead rate at MBL was about half of that at Princeton.

By the time we moved, Tsutomu was done with high school, and Sachi was in middle school. They had been experienced collectors of jellyfish, at times fireflies, and other luminescent organisms during the summers for several years, and regarded Friday Harbor as much their home as Princeton. Tsutomu, though, was particularly reluctant to leave Princeton, where he had numerous friends at the university's supercomputing center. Akemi and I had thought before about staying in Princeton and finally buying a house, and we too were sad to leave old friends behind.

In that first year, we rented a beautiful old house in Woods Hole for the winter and spring; the owners of the house, Dr. and

Mrs. Travis, would come there in the summers. Rental houses during the summer were much more expensive and harder to find because of the popularity of Cape Cod as a vacation location. So after our first summer, we knew we wanted to have our own house by the following summer.

The village of Woods Hole covers a small amount of ground, so we searched mostly in the larger Falmouth area. Finally, we purchased a wooded lot in the Sippewissett area and had a post-and-beam house built to our budget and specifications. We still live there, although the original house has been much expanded with the addition of my lab and office. When we first moved in, we found wild blueberries, delicate pink-and-green ladyslippers, and a profusion of small wildflowers growing in the shade of the oaks. We also noticed many different mushrooms in the conservation area behind our house. They intrigued me. The house of Dr. Susumu Honjo and his wife Kazuko, whom we knew at Princeton in 1963, was not very far from ours. We were surprised and pleased to find them; their presence in the area was an unexpected coincidence for us.

Our 1976 Buick parked on the street in front of the Candle House at the Marine Biological Laboratory, Woods Hole.

Lillie building, MBL; my lab was behind first-floor windows on far right.

6.3 Tsutomu and Sachi Leave Home

Although our new house included a room for Tsutomu, it was more like a guest room. In September 1982, when he was 17 years old, Tsutomu entered Caltech, and during that summer he helped a professor in Pasadena to make a computer program as he had previously promised. During the next summer, he assisted a young physicist at the Institute for Advanced Studies. In the summer of 1984, he came back to Massachusetts to work at a company near Boston, Thinking Machines Corporation (which went bankrupt or was bought by another company later), where, he told us, the renowned physicist Dr. Richard Feynman also participated in the project of creating the computer. During that time, Tsutomu occasionally came to visit us, although we seldom saw him after that; later, until about the time of the Nobel award, we could usually only get information about him through his friends.

Before Tsutomu left Caltech, we made a collection trip to the Sequoia mountains (for the millipede *Luminodesmus*), and on the way we visited the Huntington Art Gallery near Pasadena. The Huntington, surrounded by a beautiful flower garden and housing manuscripts from the Middle Ages, was a wonderful place to see. The atmosphere there was truly impressive, and we would have enjoyed seeing the collection there more often if Tsutomu had stayed longer in Pasadena.

While he was in Pasadena, Tsutomu also worked at NASA's Jet Propulsion Laboratory, and at Caltech he took only the lectures that he wanted (some of these were graduate courses with the professor's permission), while ignoring the standard procedures and requirements, much as he had done at Princeton High School. Thus, we received a notice from the university that his attendance was insufficient in regards to gym and history classes for satisfactory progress towards a degree. He quit Caltech after two years, and moved to the Los Alamos National Laboratory. This decision was another surprise to my wife and me; although by then we were accustomed to being surprised by Tsutomu, we found this an unimaginable place for him to work because of its role in the production of the atomic bomb that had so affected our lives in Japan.

Later, he moved to the San Diego Supercomputer Center. Then, on Christmas Day of 1994, the notorious hacker Kevin Mitnick, who had been sought by the FBI, invaded Tsutomu's computer and stole away some files. Tsutomu, in collaboration with the FBI, searched for and found Mitnick's residence in Raleigh, North Carolina, and succeeded in enabling his capture by the FBI. This story was reported by John Markoff in the *New York Times*, and Tsutomu suddenly became famous. We have frequently been surprised by Tsutomu's actions since his childhood; this time was the same as before.

When Sachi left for Stanford University in 1986, Akemi and I now had to live without children. Our life in Falmouth of only the two of us thus began. Because Sachi entered Stanford University, we visited the area around Palo Alto, California, twice, at her entrance and graduation ceremonies. In our visits, we saw the university campus and the surrounding area, including the Monterey Aquarium. Within the widespread Stanford campus, the garden of Rodin was our favorite place. It was quite nice to spend some quiet time with "The Thinker" in the afternoon. There are many other sculptures to look at, but the most impressive was the one titled "The Gates of Hell."

Sachi graduated Stanford University with a double major in Mathematics and English, then visited the University of Iceland as a Fulbright scholar for one year to study medieval and modern Icelandic. She was particularly interested in their sagas and manuscripts. Iceland is a small country with a population of 300,000, and Akemi and I visited once while Sachi was there. We ate delicious salmon in a lake-side restaurant in Reykjavik, the capital city, which was the only real metropolis in that country. Some of her friends took us sightseeing outside the city. Iceland is a volcanically active country like Japan with many hot springs, and with much untouched nature. I found it a very interesting country. The famous Geysir area is marked only by one or two signs and minimal rope guides to keep tourists from accidentally walking into them. We stepped back hastily when the smaller nearby geyser erupted with a tremendous burst of steam.

We didn't travel much further, but Sachi took the bus around the ring-road encircling the island, going south to Vík near the glacier atop the volcano Katla, then around the east edge of the island, and finally north with a stop at Myvatn, the aptly named "Midge Lake." It was probably well into June by then, and she could see the midnight sun just barely set under the edge of the horizon and then rise again. She met up with friends from Akureyri, the population center in the North (with less than a population of 20,000, still larger than all the other settlements outside the Reykjavik region), and they took a small ferry to an island crossing the Arctic Circle. Sachi found that the abundant tall grasses there were terrible for her allergies, but she saw puffins nesting and innumerable birds that would dive-bomb all visitors. Her Icelandic friend showed her how to wave an arm up in the air to keep the birds from diving too close, and explained that sometimes people would come gather eiderdown on the island. A famous signpost on the island indicated how far away cities like New York and Tokyo were from this isolated spot where the birds and the wind seemed the main owners.

Sachi wondered if some of the jellyfish she saw in the fjords nearby were luminescent (some of them almost certainly were, according to what she heard from the father of one of her closest Icelandic friends, a young woman who went on to become a chemist herself), and possibly even related to *Aequorea,* but I never got around to checking them out. However, later we did go on to collect red jellyfish in Bergen, Norway.

After returning from Iceland, Sachi entered a doctoral program at Cornell University in Ithaca, New York, which is relatively close to our home in Falmouth—about seven hours by car, if the weather is clear. She liked the fact that the campus was so large and sprawling there, with many undeveloped areas, similar in that way to the Stanford campus of her undergraduate years. The university is indeed surrounded by nature, located on the top of a hill above the city of Ithaca that is by the lakeside (Ithaca is in the Finger Lakes region of New York). Therefore, students must climb the hill every morning. There are rivers and falls on the campus, often very close to the stone and brick quadrangles and buildings, making a distinct contrast to Stanford University's broad campus with numerous palm trees, sandstone arches and red tile roofs. Since there are many scenic state parks nearby, we developed favorite walks during our visits; I heard that the state parks were constructed in the 1920s during the economic depression to give jobs to people. In her first year there, Sachi was dismayed to encounter snow drifts rising above her head from an unseasonable March blizzard that dumped two and a half feet of snow. Ithaca seemed at that point snowier than Iceland had been. It is a wonderful place except in the cold, icy winter, when, she said, the hill rising up to campus presents an arduous and hazardous climb.

Cornell University has a department of hotel management, with a hotel and restaurant upon campus for teaching purposes, and the orchard belonging to their College of Agriculture sells fresh fruits and dairy products. We were interested to see that their vending machines stocked their own dairy products and fruit juice. We visited each fall to pick up bushels of our favorite apples to last us for the following months. Akemi particularly loved to eat an apple with toast and tea for lunch each day.

The annual used book market put on by the Ithaca Friends of the Library book sale is well known. Inside an enormous garage, all sorts of books and phonograph records are piled up and sold. Those are materials donated in the past one year. The prices are reduced after one day, and we found that they go down to about one tenth on the fifth (last) day, to clean it all up. The parking places are very crowded. I bought many LP records there when Sachi introduced us to the sale, even though I didn't know how I would ever find the time to listen to them all. I enjoyed some of those old records upon our return home, and am looking forward to listening to more of them over the years.

Some old American universities have very fine collections of art. We often visited the art museum when we were in Princeton, alternating that trip on weekends with visits to Guyot Hall (where Tsutomu and Sachi never tired of seeing the natural science exhibits, especially the dinosaur fossils) and, on clear days, the fountains—the large Fountain of Freedom with its enormous sculpture, and the small fountain in the Prospect House garden, which had been laid out by Woodrow Wilson's wife. Akemi and I enjoyed the Rembrandt etchings of the university collection. Upon another occasion we saw an impressive prints exhibition of the Norton Simon Art collection.

Later, when we were in Ithaca, we visited the art museum at Cornell University, located atop the hill, where the city of Ithaca and the lake can be seen down below. Akemi and I always spent some quiet and satisfactory time there. Once we enjoyed an exhibition of *Jazz,* a work by Matisse; we felt lucky that it happened to be there when we visited. Apples from their orchard were another joyous gift for us in autumn.

Sachi remained in Ithaca for several years, and then taught for a year at Colgate University, about an hour and a half away by car, so we had plenty of chances to visit. Since Tsutomu had never graduated formally from his schools, and had rarely stayed in one place long enough to visit repeatedly, we were eager to return to Ithaca for Sachi's commencement ceremony when she had completed her doctorate in medieval literature. The day was sunny and the Cornell robes were a nearly fluorescent red, so the big ceremony was quite resplendent.

We also attended a separate doctoral ceremony. One of her dissertation supervisors, Professor Carol Kaske, of the same generation as Akemi and I, was interested in my experiences in Japan during and after the war, and encouraged my plans to write up my memoirs. As soon as she learned my area of research, she showed us a fascinating book, *The Rime of the Ancient Mariner,* Samuel Taylor Coleridge's well-known poem, which mentions bioluminescence at sea as an "elfish light." I was intrigued, and after returning home we arranged to obtain this book through a rare book dealer. The edition we acquired was especially interesting because it was created for the members of the Heritage Club, with illustrations by the American artist Gordon Grant; the volume, designed by Carl Purington Rollins, has its text set in Bell type and was printed on rag paper. It is a very beautiful book and we have treasured it very much since then. We were grieved to hear of Carol's passing while finishing the manuscript of this present book, in June of 2016. She was one of the people to whom we had most wanted to show my completed memoirs. When Sachi sent us the sad news that Carol had passed away, I opened my copy of *The Rime of the Ancient Mariner* and signed it in her memory.

After Tsutomu had moved to the west coast and Sachi moved to Virginia, we did not see all our family together again until my retirement symposium at MBL in 2002, just before Sachi was engaged to be married to John Brinegar. Then the next summer we all gathered together in Virginia for their wedding. It was a beautiful day for a garden wedding, before the full heat and humidity of summer set in. Our very good friends, the Ozakis, and several of our Japanese relatives were able to attend. Our relatives joked that they would all come if we could hold the ceremony in Hawaii instead.

In my MBL lab (Photo courtesy of Marine Biological Laboratory).

The following is the summary of my research at the Marine Biological Laboratory in Woods Hole during the 1980s:

6.3.1 Chemoattractant Acrasin

Long ago at Princeton, Dr. John Bonner, Dr. Johnson's close friend, came and asked me to give guidance to his postdoctoral assistant in the purification of acrasin. Acrasin, fancifully named after a character in Edmund Spenser's epic *The Faerie Queen*, is an aggregation chemoattractant of a cellular slime mold—that is, a chemical that the slime mold gives out in its single-cell stage to cause other cells to aggregate into a mobile clump of cells. I told his assistant to extract as much of the substance as possible. Without any response to my suggestion, Dr. Bonner sent a new student who asked me about the same matter, to which I gave the same suggestion. After a repetition of such things two or three times, Dr. Bonner finally lost his NSF research grant in 1980. I had a conference with Dr. Bonner about his troubles, and I agreed to purify acrasin at my lab for him if he would produce sufficient raw material. Thus, Dr. Bonner prepared the necessary raw material, which took one year, and then Akemi purified it into pure acrasin, which took a few months. The structure determination was done at Woods Hole. Since acrasin was a very simple compound derived from two amino acids, glutamic acid and ornithine, I found most of its structure by a single mass spectroscopy. Within one month I had obtained the complete structure, and informed Dr. Bonner of it.

The next day, Dr. Bonner called and jubilantly told me, "We are now toasting your success with champagne!" But I was sorry that not even the scent of champagne reached us through the phone line.

6.3.2 Luminous millipede *Luminodesmus*

We started to study a luminous millipede called *Luminodesmus* (also known as *Motyxia*) in 1980. The millipede is about 2 centimeters long, and inhabits the Sierra Nevada Mountains in California, at about an altitude of 1500 meters, and shows up on the ground under the giant Sequoia trees just after the snow has melted away, usually in the period between mid-April and early May. Since moonlight impedes locating the luminous millipedes, we chose a period of the new moon for our collection trip. We flew to San Jose in the middle of April, 1980, and drove south with a rented car about 250 kilometers to Porterville near Sequoia National Forest. Next morning, we climbed a winding road to Camp Nelson. We lost track of the number of hair-pin curves.

Finally, we arrived there. We found a small village with one motel, and two campgrounds managed by the National Park Service. It was a good area for scenic hikes. Meadows were blooming with blue lupines. Alongside the road grew tall flame-colored yucca flower spikes that fascinated my daughter.

We did not want to do our collection on private land, so we went to one of the campgrounds before dark, and waited. Fortunately, there were very few campers because it was still early in the season, and thus there was not too much light to mask the gentle glow of the worms. We saw the first blue-luminescing millipede at 8:00 p.m. when darkness came. The number of millipedes gradually increased and reached its maximum at about 9:00 p.m., and at that time we could pick up the millipedes continuously with our hands. They were most abundant under the large redwood trees. The millipede rolled up into a ball when touched by fingers, probably as a defensive reaction. When illuminated with a portable UV light, the millipede emitted intensely blue fluorescence and could be seen easily from thirty feet away. Collected millipedes were kept alive until the next morning, then anesthetized with chloroform. After removing the guts, their shells were frozen with dry-ice.

We extracted and purified the luminescent substance of *Luminodesmus* at Woods Hole. It turned out to be a luminescent protein that emitted light in the presence of ATP and magnesium ions, resembling the firefly luminescence, although it appears chemically unrelated to firefly luciferin.

Recently I saw an interesting news article. On Alcatraz Island, a small island in the San Francisco Bay several hundred kilometers northwest of Sequoia and previously well known as a jail island, which has no residents presently, the officials of the National Park Service investigated the distribution of rats that might harm birds. They spread bait marked with a fluorescent substance, intending to check rat's excrement distributions. They did not find fluorescent excrement, but instead found many strongly fluorescent *Luminodesmus* that seemed nearly identical with the ones in Sequoia, except that they were not luminous. It made me wonder: *How could millipedes move from a high mountain environment to a small island? Did they gradually lose their luminescent capability in the different environment on the island? Why might that happen?*

Left: *Luminodesmus* millipede found high in the mountains of California; Right: The luminescent worm also contains a blue fluorescent substance.

6.3.3 The use and application of aequorin

Aequorin was widely known as an excellent calcium probe by cell biologists and physiologists by 1975, and its uses and applications reached their maximum around 1985. We were still practically the only source of purified aequorin, so we received requests from investigators from all over the world. As a result, we mailed out several hundred aequorin samples. Although the cDNA of aequorin was cloned and recombinant aequorin was created in 1985, the patent owners of recombinant aequorin, the University of Georgia and Chisso Corporation of Japan, made no cooperative efforts to expand the use of aequorin; thus recombinant aequorin did not become generally available and used for some time. On the research side, we collaborated with Professor Yoshito Kishi of Harvard University to make various aequorins possessing different calcium sensitivities and properties. Those aequorins were called semi-synthetic aequorins, and were prepared by replacing the coelenterazine moiety of aequorin with a derivative of coelenterazine. The replacement of this part of the molecule caused some of these semi-synthetic aequorins to have more sensitivity or less sensitivity to calcium, so that researchers in different areas could choose the one most suitable for their research. Their applications further expanded the uses of aequorin as a calcium probe for research and experiments in physiology.

Akemi, working as my lab assistant, took on most of the job of purification. Aequorin and GFP purification were almost year-around work for her. Later, she would also work on the fungus experiments with me and purify luciferase from the reddish Norwegian jellyfish *Periphylla*.

6.3.4 Luminous brittle star

We studied the luminous brittle star *Ophiopsila* in collaboration with Dr. Paul Brehm, a professor at the State University of New York at Stony Brook at that time. *Ophiopsila* is a brownish brittle star, with five snaking thin arms of 5 centimeters long, and abundantly seen on the shores of Catalina Island near Los Angeles. We extracted and purified the luminescent substance of *Ophiopsila*, and found that it was a photoprotein that emitted light in the presence of hydrogen peroxide. I later learned that it was through a seminar given by Paul Brehm that my co-laureate Martin Chalfie was inspired to use GFP in his own research.

6.4 My Friend Goto

Lastly, I would like to mention my sincere friend, the late Professor Toshio Goto, who was only one year younger than me and was, like me, a graduate of a pharmacy college. He had worked with me in the purification and structure study of *Cypridina* luciferin since 1956, as instructed by Professor Hirata. We collected *Cypridina* together on the

Chita peninsula, and developed a close relationship. After my move to the United States, he visited me almost every year. Even my children looked forward to Goto-san's arrival. I always saw him when I visited Nagoya. Once when he visited our Princeton apartment, he was interested in our Steinway piano (which I had bought as a gift for Akemi), and trying it out briefly, he said, "It's not much different from my Yamaha."

When he came to Friday Harbor, we fished together from the rocky shore, and he enjoyed the strong pull of the black rockfish on the line. Then we made sashimi, and invited Dr. Johnson. I also met him several times on trips; at the University of Georgia in 1972, and in London in 1975 when I was invited by one college in London and he was invited by another college. In 1978, while we were both attending a symposium in a seashore town in Brazil, Professor Goto stayed in the next room. I met him for the last time in 1990, when Professor Kishi of Harvard University invited me to Boston for a dinner when Professor Goto visited him. A few months later, Goto died suddenly, and I lost my oldest friend. He was 61 years old. Goto was a superb scholar and educator.

CHAPTER 7

The 1990s: Losses, Gains, and the Rise of GFP

7.1 Dr. Johnson's Passing

The 1990s were to bring further losses. Only one month after Dr. Goto's death, in September 1990, my long-time mentor and friend Dr. Frank H. Johnson passed away at the age of 82. Several years before, he had had a stroke. I felt sorry that I had not gone to Princeton to see him during that period. In October 1990, Akemi and I attended his memorial service, held at the Princeton University chapel. Mrs. Mary Johnson and his three daughters, Lanie Johnson, Frances J. Cunningham and Charlie Pierce, were very glad to see us and welcomed us enthusiastically. At the service, Dr. Johnson's friend Professor John Bonner and Lanie delivered speeches. Their speeches, together with the chorus sung by his three daughters and his grandchild's Bible reading, made the service memorable and wonderful. Dr. Johnson, with his deep dedication to bioluminescence study, would have appreciated the idea that they lit up the darkness of our grief. I was also very glad to see another of Dr. Johnson's friends, Dr. Aurin M. Chase, a Professor of Biology at Princeton, who was another student of Newton Harvey. All my dried *Cypridina* samples at hand now were inherited from Professor Chase.

Dr. Johnson's passing marks the end of an era for me—an era that began in 1955 with my fortuitous study of *Cypridina*. As the third of my mentors, he was instrumental in my move to the United States. He was born in 1908 in North Carolina, and graduated from Princeton University in 1930. Having received his Master of Science at Duke in 1932, then a PhD at Princeton in 1936, he was a true Princetonian and inherited Newton Harvey's bioluminescence work there. For many years, he taught microbiology at Princeton. His best friend was Henry Eyring, a theoretical chemist at the University of Utah; they published two books based on their collaborations. His favorite hobby was drawing and painting. He particularly liked to sketch female figures, and many of those were hanging in his house. He had wanted to draw my portrait, but the project had never materialized. Two of his paintings still hang in my house. He once told me that he might be more successful as an artist than a scientist, but I am not so sure.

In research, Dr. Johnson used to discuss everything with me on equal terms, but after the aequorin extraction he avoided discussing research as much with me, and did not wish to oppose my ideas at all. That meant I had to be more careful when I talked to him. For fifteen years until his retirement he did his best to help me to make advances in bioluminescence research. He was extremely skilled at making almost anything; he made the first crucial jellyfish cutting machine, which helped bring about the successes in aequorin research. He did not teach me very much in science, since much of my scientific learning was through trial and effort, and attention to details that could eventually show me the right way to proceed, as in my first work with *Cypridina*, but one important thing that he taught me, which I will never forget, was that when you have questions or even a little doubt about words in your writing, you should always consult with dictionaries. He himself did so all the time. I always thought of him as a typical gentleman, and he was one of the three people (along with Professor Yasunaga and Professor Hirata) whom I respected most highly in my life.

7.2 The Passing of Relatives in Japan

In January 1991 my younger sister Setsuko Ito died in a car accident. Akemi and I were shocked by the unexpected tragedy. She was just 47. Her three children were close in age to Sachi. She had been taking care of our mother, and I am afraid and am sorry that she might have been overworked by the care necessary to tend an elderly person. If I had lived in Japan, then my wife and I would traditionally have borne the responsibility for my mother's care, since I was the eldest son. Setsuko and her husband Mr. Kanae Ito were very kind to me and helped me a lot, not only in taking care of my mother, but also in many other ways.

In the same year, in August, my brother-in-law Yoshio Yamane, MD, passed away of lung cancer. He was the same age as me, and we had been friends for a long time, ever since my marriage with Akemi. He had been one of the closest friends of Akemi's elder brother since kindergarten, so that he had known Akemi's mother well even before Akemi was born. His father was a professor of Nagasaki University Medical School, but he applied to the Japanese navy, unlike Akemi's brother. Thus, ironically, his affiliation with the navy enabled him to survive the atomic bomb in Nagasaki, which had killed so many of the medical students and faculty. His father was a victim of the bomb, like Akemi's brother. After the war he decided to study medicine like his father. He was Akemi's older sister's husband. Every time I visited Nagasaki, he entertained me with delicious Japanese food. So in just one year, I lost two persons who were important to me.

Two years later, in 1993, my mother passed away in Isahaya. She was 92. For two years after my sister's death, my brother Sadamu and his wife Akiko Fujiyama had taken care of my mother in the hospital. As a wife of a military man during World War II, my mother's life must not have been an easy one, but she had always done her best to keep her family happy. The next year, Akemi's mother passed away at age 89. She was very kind and possessed an unusual amount of thoughtful consideration, like my grandmother Tsuki. I had always stayed at her house when visiting Nagasaki. She was very popular because of her kindness to everyone; many people missed her at her funeral. Again, an era had ended for Akemi and me. I went back to Japan for her memorial service a year later with Akemi and our daughter Sachi.

My research also reached an unexpected turning point around this time, as I will explain.

7.2.1 Luminous mushrooms

When I was fifty-seven years old, I privately planned that I would stop my study of bioluminescence fifteen years later in the year 2000, because I wanted to write and publish a book on bioluminescence for future generations of researchers while I had the energy to do it. Thus, I chose the study of luminous mushrooms as my last major project.

Luminous mushrooms (fungi) have attracted a wide interest and have been studied by a great number of scholars since ancient times. About forty genera of luminous mushrooms are known, and all of them emit almost identical greenish light. As its only chemical study, Kuwabara and Wassink studied the luminous fungus of Puerto Rican coffee leaves in 1966. They cultivated the fungus and obtained 15 kilograms of mycelia, from which they extracted and purified a luciferin; however, they achieved no conclusive result. No chemical investigation had been attempted since then, probably due to the apparent chemical complexity of the luminescence system. I suppose everybody avoids studying a troublesome matter that would likely fail to yield results. Indeed, fungus luminescence is highly complicated. The light emission takes place only during developmental stages specific to the species. Moreover, sometimes luminous species change into non-luminous species for no obvious reason, and vice versa.

One reason I chose to study mushrooms was because we had found so many species growing near our house. I had become curious about them. We discovered that the oak logs left from clearing land for our house would readily grow both wild and cultivated kinds of fungi, so we had cultivated our own shiitake mushrooms. We saw mushrooms everywhere as we tended our property or walked in the nearby wooded areas; some, I learned, were known to be luminous.

First, I studied five species of luminous fungi in different genera, measuring such aspects as the amount of luciferin, the intensity of luminescence, and the amount of superoxide dismutase (SOD) that existed during the course of the development of the fungi. I arrived at the conclusion that superoxide is needed for the luminescence of fungi. Superoxide is an unstable free radical that is situated between oxygen and hydrogen peroxide.

The luciferin in luminous fungi is continuously synthesized as it is consumed by luminescence, just like the aldehyde (luciferin) in luminous bacteria. Because the amount of luciferin in fungus would be very small and difficult to study, we decided to look for the precursor of luciferin. We found a high content of this precursor in the fungus *Panellus*, which has an interesting nature; its luminescence is extinguished by dry weather and restored by rain. We extracted and purified the precursor, and found its chemical structure. We also extracted and purified a luciferin precursor of another fungus, *Mycena citricolor*.

Left, above: *Panellus*, a common luminous mushroom often seen on oak trees broken by storms; below: *Panellus* in darkness. Right: oak log segments left under the trees off our driveway to grow fungi.

However, my research was interrupted when I confronted a problem in 1995 concerning aequorin, as explained below.

7.2.2 The X-ray structure of aequorin

The outline of the structure of aequorin and the mechanism of its reaction were understood in 1979, when the presence of coelenterazine peroxide in the aequorin molecule was concluded. However, the complete structure of the aequorin molecule had to wait on the results of X-ray diffraction analysis of aequorin crystals. Although natural aequorin is heterogeneous and not suitable for X-ray analysis, recombinant aequorin is homogeneous and suited for the purpose. Around 1990, three different groups began X-ray analysis of recombinant aequorin. One of the groups sent me a signed framed photograph of the diffraction pattern of aequorin. I interpreted it as an expression of their confidence. I thought that group, as well as the other groups, would soon succeed in the X-ray analysis. I waited eagerly to hear about their success, but then I heard a rumor that all of them had failed. It was sometime in early 1995.

I thought that, if they had failed in the X-ray analysis of aequorin, the cause must be the presence of impurities in the aequorin. I was confident that I could make purer recombinant aequorin because I knew aequorin better than anyone else. I was faced with making a decision whether to continue my work on luminous fungi, or to change my work to the X-ray analysis of aequorin. I chose the X-ray study of aequorin, giving up the study of luminous fungi, because aequorin had been the most important work in my life.

I first devised a new purification method of aequorin. The method resembled the Anderson method that I mentioned in the purification of *Cypridina* luciferin, in changing the subject substance temporarily into a derivative having altered properties: in this case, apoaequorin. (Aequorin and apoaequorin have this difference in chemical properties: when a sample is adsorbed on the hydrophobic interaction column of butyl sepharose 4 in 2 M ammonium sulfate, and the column is eluted with decreasing concentrations of ammonium sulfate, aequorin is eluted with 1 M ammonium sulfate but apoaequorin can be eluted only with less than 0.2 M ammonium sulfate.) By temporary conversion of aequorin to apoaequorin, therefore, it is possible, at least theoretically, to obtain 100%-pure aequorin.

Then, I organized a team. Dr. Teranishi of Mie University synthesized coelenterazine and its derivatives. Dr. Inouye of Chisso Corporation prepared *E. coli* carrying the apoaequorin gene. My job was mass cultivation of the *E. coli*, extraction and purification of apoaequorin from the mass culture, and then the conversion of apoaequorin into aequorin, and finally the purification of aequorin. Dr. James Head of Boston University would complete the crystallization and X-ray analysis. In carrying out my part, however, I had to learn the methods of bacterial cultivation from textbooks because I had no experience of bacteriology.

We had unexpected trouble at the end of this project. Dr. Head called and informed me that there was no peroxide group in the molecule according to his interpretation of the electron density map he obtained from X-ray diffraction data of aequorin. I was confident that aequorin contained coelenterazine peroxide based on my chemical data as well as NMR data obtained later. I visited his lab in Boston and discussed the data with him, but I could not convince him, and he insisted on the absence of a peroxide. Finally, I gave up and told him, "We should report what you have found, but please do not conclude the absence of peroxide," and I left. One week later, he called me and said, "There was a peroxide group. The problem was that its electron density was unusually weak." I was greatly relieved. It was the end of 1999.

The three-dimensional structure of aequorin obtained by X-ray diffraction was published in 2000 in *Nature*. It indicated that aequorin was a globular protein containing coelenterazine peroxide (the light-emitting group) at the center of the molecule. Thus, the light-emitting group is covered with a layer of protein, obstructing reactions from outside. This explains why we had been unable to extract the light-emitting group from aequorin. By the way, the binding of calcium ions to aequorin causes structural changes to the protein part of aequorin, which destabilizes the light-emitting group, causing its decomposition to coelenteramide and carbon dioxides accompanied by light emission.

7.2.3 Developments in the study of GFP

The cDNA of GFP was cloned by Dr. Douglas Prasher in 1992. It was thought at that time that expressing the cDNA in living organisms would not actually produce fluorescent GFP, because specific enzymes for the required condensation and dehydrogenation reactions would be needed to form the fluorescence chromophore. Two years later, in 1994, Dr. Martin Chalfie of Columbia University overturned this assumption. Upon attempting to express the cDNA in *E. coli* and a nematode worm, he observed the unexpected fluorescence of the GFP, showing that the necessary condensation and dehydrogenation reactions must have occurred automatically and spontaneously. His findings suggested that it was in fact possible to express GFP (and other proteins bound with GFP) in living organisms to observe their internal processes or behavior. This ground-breaking work by Dr. Chalfie attracted widespread interest and launched development of further applications of GFP.

GFP became well known to the general public by the appearance of fluorescent animals, such as a Medaka fish, mouse, frog, and rabbit. The most famous of them was probably the rabbit named Alba that was created at a Paris laboratory by order of a Chicago artist who wanted to display the animal as an object of art. However, it brought on public criticism, and the laboratory refused to hand the animal over to the artist. Fluorescent Medaka fish and zebra fish are produced in Taiwan and being sold widely. There are various problems in producing such fluorescent animals, including obvious ethical issues, and it seems doubtful that this application will contribute much to the welfare of mankind or the progress of science.

GFP has become an essential tool in the fields of biology, physiology and medicine, especially as a fluorescent marker of protein molecules and cells. As its use and applications have widened, various improvements have been sought in the color of the fluorescence and the properties of the protein. Because a red color suffers less from scattering than blue and green colors for use in experiments in cells and tissues, many investigators rushed to find red fluorescent proteins, by diving into the sea. However, the first red fluorescent protein was found by Russian scientists in 1999, ironically from a coral purchased at a pet shop.

In the study and improvements of GFP, we cannot forget the great contribution made by my esteemed colleague in GFP studies, Dr. Roger Tsien. He prepared many different fluorescent proteins to emit various colors of fluorescence, from blue to red, by modifying the amino acid residues surrounding GFP's chromophore. In fields outside of biology and medicine, GFP plays a role as a sensor to detect substances such as cadmium and zinc, or even the explosive TNT; the potential uses of the fluorescent proteins now appear beyond our imagination.

7.3 The Value of Pure Science

It is important to me to remember (and to remind all students of science) that such directions of research had, as their precursor, a study of purely scientific matters. I had no notion of the possible applications when first isolating and purifying luminescent proteins and GFP. But no applications are possible without first understanding the fundamental chemical details of substances in nature. Research for the sake of knowledge continues to be crucial, particularly since we often cannot know the importance of chance discoveries until many decades later.

To reminisce, an article "Bioluminescence: A Princeton Discipline" by Laurence B. Chase in the *Princeton Alumni Weekly* (volume 67), mentions, in 1967, some proposed uses for bioluminescence, including in outer space: since firefly luciferin is triggered by ATP, a necessary molecule for living organisms, that perhaps it could detect alien life on other planets; or that other luminescence systems that dim in the presence of contamination by fuel residues could be used to detect atmospheric traces left by space vehicles. Such ideas may seem farfetched, but perhaps no more farfetched than the applications that have arisen from aequorin and GFP research and become accepted as a standard part of science today.

7.4 Norway and the Jellyfish *Periphylla*

While my daughter was in Iceland on a Fulbright fellowship in 1992, she travelled to visit some friends at their homes in the north of the island, and reported seeing remains of jellyfish that resembled *Aequorea* on the edges of some of the fjords. She wondered if they might be luminescent, and she had also heard of some marine luminescence that fishermen had noticed. We visited Sachi in Reykjavik, and met her friends. One of them, Vigdis, an adult student, kindly offered to drive us on a daylong tour of the major nearby sights: the geysers, Gullfoss (the "golden waterfall"), and the ancient location of the Icelandic parliament, the Althing. She and other friends, some of whom were studying fisheries, biology, chemistry, or literature and languages, talked about some of the fascinating fauna of Iceland. (I heard that her friend Rut Valgardsdottir went on to use GFP in her own research, some years before my Nobel Prize.) The salmon in Reykjavik was delicious, too. So I thought that I might study some northern luminescent species if the opportunity arose.

The opportunity eventually arose. The bioluminescent jellyfish *Periphylla* is widely distributed in the world. Its size is usually several centimeters in diameter, not so different from *Aequorea*. In certain Norwegian fjords, however, they grow to over eight inches in diameter, weighing nearly two pounds. Because the entrances of fjords are narrow and shallow, their environment probably differs significantly from the open ocean. I began to study this jellyfish in 1996 using specimens given me by Dr. Par Flood of Norway, and I quickly found that its luminescence was caused by coelenterazine and a luciferase. This luciferase, however, was an interesting enzyme with an unusually high activity level.

Shortly after, Akemi and I had an opportunity to study the jellyfish on the research ship *Hakon Mosby* of the University of Bergen for one week, and I was able to extract *Periphylla* luciferase onboard. This was my first time aboard a research vessel, and we both had a very interesting experience. The *Hakon Mosby* was a 300-ton ship designed to accommodate ten researchers. I took all my chemicals and chromatography instruments as well as dry ice with me. Upon boarding the ship at Bergen, first we needed to go through training for emergency situations: We had to put on heavy rubber suits and sit in a lifeboat. (The water temperature was 12 degrees Celsius; one could survive only for thirty minutes without the suit's protection.) It was not an easy thing to do, especially for Akemi, because the suit, made for six-foot-tall men, weighed about forty-five pounds, while she was a typically petite Japanese woman, only just over five feet tall. Once our suits were on, we had to run and sit in a lifeboat that could accommodate twenty people and which was completely sealed. The emergency training was a difficult undertaking for older persons, like myself, and for petite women. Aside from those things, it was a very pleasant and interesting trip, with good food to enjoy and plenty of opportunities to admire the beautiful scenery of the fjord.

Research ship *Hakon Mosby*, University of Bergen.

***Periphylla* from fjord in Norway (this specimen is 750 grams and about 20 years old).**

Periphylla was collected with a net of over six feet in diameter, which could obtain about 100 kilograms of specimens from a depth of about 200 meters, with the aid of sonar; it was a striking contrast to the methods we used to collect the jellyfish *Aequorea*. Then I extracted the luciferase from the upper surface of the umbrella and from the ovary. The luciferase from the umbrella was resistant to heat, and the one in the ovary was resistant to acid and alkali. It seemed interesting, although the reasons why the luciferase differs in these ways remains unknown. So much of scientific study consists of small, often mysterious, chance discoveries that may someday combine with other miscellaneous findings to create a new area of research and to yield previously unimagined applications.

We joked that the jellyfish were shaped like large red breasts. So Akemi, as a jest, held up two of those large jellyfish in front of her chest, and we took a picture. The *Periphylla* showed very vivid and bright against her teal-green parka.

We made two trips to Bergen (in 1998 and 2000) to collect *Periphylla*, and stayed each time at a hotel near the harbor. The city of Bergen is surrounded by hills, like the city of Nagasaki. We saw the famous fish market under the windows of our fifth-floor room. Beyond the harbor, several Bryggen-type historical buildings were set on the shore. These are large storage buildings, preserved or restored as remains from the Hanseatic-League era, and now the first floors are being used as gift shops. It reminded me of the Dejima in Nagasaki.

We experienced the fact that salmon and potato are very common foods here, and the dishes we ate at the Hotel Norge were really good—we enjoyed the buffet-style spread with many kinds of fine dishes, some of regular western style, others traditional Norwegian or maybe Viking-inspired too, so many that we could not try them all. Also I visited a restaurant near the fish market that served Louisiana-style foods, with my friend Bruce Bryan, MD, who has a strong interest in bioluminescence and holds a patent on bioluminescent novelty items. The seafood served there was very delicious. I would not mind visiting Bergen many times.

CHAPTER 8

Going Full Circle: Nature's Gift, Scientific Honors, and Revisiting Japan

My National Science Foundation grant ended in the year 2000; the following year I retired from the Marine Biological Laboratory. As my plan was to continue performing some experiments, and I was still receiving occasional requests for aequorin samples, we set up lab space at home and moved all of my equipment and chemicals there, with permission from MBL. The permission was granted because all of them had been purchased with my grant and, moreover, all of the equipment was old enough to have exceeded its usual duration of usage. (Some of it had even accompanied me from Princeton University, two decades earlier.) It would have cost MBL more to dispose of them than to let me take them.

8.1 After My Retirement

On July 27, 2002, my retirement symposium, titled "GFP and Aequorin," took place in the Lillie Auditorium of MBL, by the thoughtful arrangements kindly made by Dr. Shinya Inoue, a worldwide authority on microscopy and creator of the Shinya Scope. Famous people in my field, such as Martin Chalfie, Roger Tsien, and

William Ward, joined the symposium from distant places. My children were both able to visit at the same time, for the first time in several years.

At about this point, I started to write a book on bioluminescence, which I had been planning for a long time. It took three years to finish a draft version, and then I began to look for a publisher. Princeton University Press and John Wiley did not respond at all to my inquiries. I thought that they probably had no interest in such a narrow field as bioluminescence. Then I happened to see an advertisement for World Scientific, so I inquired with them about the possibility of publishing my book. The editors responded immediately, saying, "We would like to publish your book, but please ask a native English-speaking person to take a look at the manuscript." I asked three people, including Dr. Steven Haddock of the Monterey Bay Aquarium Research Institute, to fix errors in my English, and also asked three other people for their comments on the contents. Finally, after three stages of proofreading and innumerable small corrections, my book was published in September 2006. About 500 pages long, it discusses the known chemical principles and experimental methods for about thirty-five kinds of luminous organisms that had been chemically investigated.

I wrote this book especially for future students who might study the chemical aspects of bioluminescence. Because I have had the broadest experience in the chemical studies of different bioluminescent organisms, I felt that writing this book to convey the range of methods that have been developed to research and understand bioluminescence systems was my particular duty to future students. It perhaps repays a portion of my debt of gratitude to the mentors who encouraged me to travel my chosen paths in science, no matter how uncertain or unseen the outcomes of those paths. They shared their knowledge and experience, inspiring me to explore previously unmapped side roads. I hope my book will likewise inspire others to devise methods to learn about further aspects of bioluminescence, or other biological systems, whose fascinating chemistry is currently undreamt of.

The summer of 2004 was my most splendid summer. I received the Pearse Prize from the Royal Microscopical Society in July. The awarding of the Pearse Prize, which honors significant strides in life sciences (and in research to identify and understand the chemicals involved with living cells and tissues), reflected the growth of applications for bioluminescence and GFP as markers or tracers within living cells and organisms. For instance, GFP could be used to make the spreading of cancer cells visible; aequorin could be used to show changes in calcium levels during a process, like cell division, within living cells or organisms. The Pearse Prize Lecture Session was held at the Twelfth International Congress of Histochemistry and Cytochemistry (ICHC) meeting at the University of California in San Diego. The title of my lecture was "The discovery of aequorin and green fluorescent protein." Next, I was invited to the International Bioluminescence and Chemiluminescence Symposium at Yokohama, Japan in early August and gave a special lecture titled "Aequorin & GFP: An historical account."

A special event marked the end of August: I attended the symposium "Calcium-Regulated Photoprotein and Green Fluorescent Protein" held to honor me at the Friday Harbor Laboratories as a part of the centennial celebration of the laboratory. Most of the well-known researchers from related fields gathered there at the Friday Harbor symposium, arriving from all over the world. These included Dr. Atsushi Miyawaki and Dr. Takeharu Nagai and a few more scientists from Japan, as well as my co-laureates Chalfie and Tsien. The picnic party held at the beach of retired Professor Arthur Whiteley's property at Friday Harbor was particularly memorable for participants. It was a beautiful day and we enjoyed the splendid view. Dr. Whiteley's property faced towards the mainland, so that we could see the mountains far away against the blue sky, behind the deeper blue water and the green islands. On a clear day Mt. Rainier could be seen there, its snowy peak floating in the distance. Shinya Inoue, Martin Chalfie, Roger Tsien, William Ward, J.R. Blinks, Claudia Mills, Lionel Jaffe, the group of Japanese scientists from the research institute Riken, and Akemi and I were among more than thirty participants.

Claudia Mills, who was a research scientist at Friday Harbor Laboratories for over three decades, kept the jellyfish collecting nets under her custody. In years past, when we traveled to Friday Harbor to collect jellyfish almost annually, we would often store them, wrapped up in brown paper, in the rafters of the Gear Locker between summers. She displayed one of these nets in front of the lecture room during the meeting. She also helped to arrange to send one of these nets to the Nobel Museum in Stockholm. I think Dr. Johnson, who painted the handles of the nets in the orange and black of Princeton University's colors, would have been pleased but surprised at having his nets displayed there!

It was a very special reunion. Friday Harbor had become our summer home over the many years of aequorin collection. I was glad to see so many of my old friends, but I was also saddened to learn that as the town's tourism had expanded tremendously over the years, the sea at Friday Harbor was steadily being polluted.

8.2 Nature's Gift: Abundance and Disappearance of *Aequorea*

I would like to add a bittersweet mention here regarding *Aequorea*. We traveled to Friday Harbor nineteen times to collect the jellyfish from 1961 until 1988, obtaining a total of about 850,000 specimens of *Aequorea*. We made thirteen of these trips in our own car, driving over 6000 miles round trip each time. Later, we would fly, then rent a car in Seattle—one with a trunk large enough to hold a dozen buckets of jellyfish. By the mid-1970s and later, my entire family was very experienced at collecting jellyfish and cleaning and repairing nets. We would spend most of two months each summer collecting them, counting them into buckets, and cutting their edges in the mornings, and sometimes, if the tides ran full of jellyfish, into the evenings, too. Then we would extract the luminescent substances every afternoon out of batches of jellyfish rims (which we called "rings"). Possibly, we would even dream about streams of *Aequorea* at night. Our hands smelled constantly like jellyfish during those weeks. Our "jellyfish factory" ran smoothly most years, and we enjoyed the natural beauty and sea air at Friday Harbor. On weekends, we would fish off the rocky shores, collect abalone, trek to our favorite locations around the island, or check out a boat from the lab dock. We would fetch groceries and a cake from town, or buy a salmon from a fisherman, repair torn nets, and enjoy watching the ferries and the teeming life of the shore and the sea.

The jellyfish were always abundant at Friday Harbor during that period, although they sometimes disappeared for a few days, depending on the direction of the wind.

We would walk along the docks and easily find clusters of *Aequorea* and a few other jellyfish species in the gaps between docked boats: *Halistaura* (now called *Mitrocoma*), which was a flat transparent pancake with a pair of crossed lines through the center; tiny marble-sized *Stomatocha*, which my daughter thought looked like clear elf hats with a solid spot in the middles; or iridescent comb jellies with long ciliated tentacles curving through the water. Occasionally we would see larger, colorful lion's mane jellyfish, vivid reddish-orange or yellow, whose stinging tentacles we did not want to tangle onto our collecting nets. Sometimes the currents even brought what seemed like an almost solid clump or continuous stream of jellyfish on the ocean surface, covering an area a meter or two wide.

We made our last collection trip in 1988, when my younger child was at Stanford University. I remember *Aequorea* was very abundant that year, facilitating our collecting some 100,000 specimens. Mysteriously, however, the jellyfish population decreased drastically since then, so that they were rarely seen after 1990, and we experienced difficulty in collecting even a few specimens. This depletion of the jellyfish population was not a result of scientific study. We had rarely collected jellyfish smaller than about 7–8 centimeters in diameter for our research, since these were too small to process efficiently. Also, we and other scientists who studied jellyfish regularly witnessed so many washing by on the currents that we could only collect a small fraction of the larger ones. The sudden drastic decrease may have resulted from the environmental contamination of the sea bed by crude oil spilled by the Exxon tanker *Valdez* near Alaskan shores in 1989, or from the warming of the earth, or it could have resulted from other causes yet unknown.

Our study of *Aequorea* was extremely lucky in its timing; it seems we did our study by the direction of God. If we had started just twenty years later, GFP would be unknown today. Because recombinant aequorin became obtainable in 1985 through successes in cloning, we no longer needed natural aequorin, either as a calcium probe or to further study the properties of aequorin. Therefore, we no longer needed the jellyfish *Aequorea*, at least from the viewpoint of biochemical discovery. That is, we did not need to collect more jellyfish because we could now obtain both aequorin and GFP more efficiently by cloning. We had completed all of the necessary analysis just before it became impossible to study the biochemistry of the jellyfish in its natural form any further. It was as if we had been guided by the heavens.

Although we discovered GFP in 1962, it was purified as a serendipitous byproduct of aequorin, as I have already mentioned. Its discovery was apparently accidental. However, GFP is very widely used now as an indispensable protein marker, due to the fact that the peptide chain of GFP self-contains the chromophore, enabling the expression of GFP in living bodies.

Could anyone have predicted the existence of a protein that self-contains a chromophore? Without that knowledge, it would have seemed very unlikely for anyone to begin a study that required so many thousands of jellyfish, and consequently GFP would most likely never have been discovered purposefully. In our case, however, we needed aequorin in order to study its luminescence mechanism, and also, after discovering its sensitivity to calcium ions, to supply aequorin to biologists and physiologists who needed it in their research. All of this required many thousands of the jellyfish. Consequently, as a side result, we obtained sufficient GFP to study its chemical properties and identify its unusual structure.

Many circumstances intersected to give an unexpected impact to our research. Because I had begun my research career by pursuing paths with little likelihood of quick results and no clear protocols for successful completion, such as crystallizing *Cypridina* luciferin at the beginning of my career, or searching for an unknown substance, rather than the known substance luciferin, in *Aequorea*, I was willing to persist in a long-term study of 850,000 jellyfish. But I might still not have learned from chance results, like the unexpected flash of light when aequorin solution flowed into seawater. I might not have understood how much light the study of one biological system could shed on a different system. Science enables us to put together these strands of knowledge, thread-like as jellyfish tentacles.

Such scientific discoveries arise from the ecosystems of ocean or forest whose preservation today seems increasingly delicate, as humans continue polluting places of former paradise. Upon considering those matters, the discovery of GFP and its development to its present status seem to me the results of an act of God.

Watching a ferry pass by (Friday Harbor Laboratories, 2004); photo provided by Dr. Takeharu Nagai.

Working in the lab set up within my house (Falmouth, 2005).

8.3 More Honors, The Asahi Prize, and Its Influence

In the summer of 2005, I was awarded the Émile Chamot award from the State Microscopical Society of Illinois. I traveled to Chicago to receive the award. It had been 44 years since my last visit there. When I gave my lecture about the discovery of aequorin and GFP at the Inter/ Micro 2005 symposium (an international microscopy conference) held at the McCrone Research Institute in Chicago, Illinois, everybody in the meeting was surprised or amused to hear about my visit to Chicago so long ago and our old experiences in the city. I visited the Art Institute of Chicago again as I did in 1961, and enjoyed strolling and window shopping on Michigan Avenue. I enjoyed a delicious lunch at the Four Seasons Hotel restaurant.

The city of Chicago has obviously changed greatly over the forty years; it seemed much cleaner and brighter, and more full of people. This visit to Chicago provided me and Akemi with wonderful opportunities to locate art galleries whose names we knew, but which we had never found the chance to visit before. We enjoyed seeing many prints, those from both old masters and from modern artists whom we admired very much.

Unexpectedly, in 2006, I received the Asahi Prize, a major prize in Japan honoring work in academic and artistic fields, from the *Asahi Shimbun* Company for my "discovery of green fluorescent protein and its contribution to life science." *(Asahi Shimbun* is a prominent daily newspaper founded in Osaka in 1879; its awards have been given since its fiftieth anniversary in 1929.) The award ceremony took place on January 29, 2007, at the Imperial Hotel in Tokyo, and many of my relatives and friends whom I hadn't seen for many years, including my son who was now based in California and Oregon, were invited. I was very happy to see over fifty of my relatives and friends. Such prizes are usually given through someone's strong recommendation, which I would have expected to be unlikely in my case, since I had worked in the United States for so many years that my name would have been almost unknown in Japan. I suspect that the prize was awarded to me by the recommendation of Professor Issei Mabuchi of Tokyo University (presently at Gakushuin University) who visited MBL every summer to do research.

On the morning of the award ceremony, I received one box of Mikan oranges at the Imperial Hotel where I stayed. I wondered what to do with such a large quantity of oranges. However, instead of oranges, I found in the box a heavy bronze bell, 16 centimeters in diameter, apparently a reproduction of the Angeles Bell of the destroyed Urakami cathedral, which had survived the atomic bomb near ground zero and is one of the symbols of Nagasaki. An attached note from Mr. Tadami Mine indicated that it was a gift from my classmates at the Nagasaki Pharmacy College. The bell now sits in the entrance room of our home.

The Asahi Prize gave me an additional benefit. In Japan, "aequorin" was usually pronounced "equorin," and I had tried unsuccessfully for almost ten years to correct the pronunciation to "iquorin." However, the president of the Asahi Company, Mr. Akiyama, declared in his speech at the ceremony that the use by Asahi of "equorin" was incorrect, and should be changed to "iquorin," in front of an audience that included many researchers in bioluminescence. Since then, it has become rare to hear the incorrect pronunciation.

One week after accepting the Asahi Prize, I was invited to give a talk at the third Hirata symposium at Nagoya University on February 6. This invitation was very welcome, since I had been hoping to attend the symposium, which had been established in memory of the late Professor Yoshimasa Hirata a few years before. It was another benefit emerging from my Asahi Prize. The lecturers were invited for dinner on the previous evening at the Marriot Hotel at Nagoya railway station. I arrived about thirty minutes late because I was unfamiliar with the location of the hotel; it was a small reminder of how much Nagoya had changed since I had left in 1965, but the dinner gave me a welcome opportunity to meet old friends from the Hirata lab after so many years. I enjoyed a very pleasant evening.

On October 20, 2007, both I and Dr. Ryuji Kobayashi of the University of Texas, who was also a graduate of the Pharmacy Department of Nagasaki University, gave talks at the Nakabe Hall of Nagasaki University. On that occasion, I received the title of honorary alumnus of Nagasaki University. As the third recipient of that title, I was also the first scientist to receive it. It was apparently another effect of the Asahi Prize, which had rendered me much more visible in Japan. The fourth recipient of that honorary title was Mr. Shigeo Fukuchi, the chairman of NHK (Japan Broadcasting Company).

I found out that Dr. Kobayashi, who gave a talk at Nagasaki University on the same day as I, had been born in the town of Kojiro, near Saigo, my permanent domicile and the place of my family registry, in the Shimabara peninsula. His grandmother, Kuma Kobayashi of Kobayashi Pharmacy, had been a friend of my grandmother, Tsuki Shimomura; I remember visiting Kobayashi Pharmacy when I was a child, accompanying my grandmother. Sometimes the world is an unexpectedly small place.

I had never returned long-term to Japan, since 1965, in the course of my career. Akemi and I had long since put aside the idea of "when we go back to Japan" as a wistful dream; we knew that we would never go back to stay. However, with the recognition I received for my work, well after my retirement, I was visiting Japan more often than I had done for many years. These visits were to occur even more regularly after the Nobel Prize. But if I had not decided to leave my prospective future in Japan so many decades ago, I would not have been able to pursue my bioluminescence research so intensively. Ironically, my decision to leave Japan had now brought me back; my life's path seemed to be tracing some sort of large circle.

It appears that all the lectures I gave and the awards I received since 2000 were contingent on the discovery of GFP, which was incidental to my main field of research, bioluminescence. It makes me sorry in a sense; it is somewhat disappointing to me that my attempts to open up the field of bioluminescence have been eclipsed by the prominence of GFP after I had specialized in the chemical research of bioluminescence and had made it my life's work for half a century. Yet this is the reality of the world in which I live.

CHAPTER 9

The End of the Path: The Nobel Prize Announcement and Nobel Week

At the end of September 2004, Fuji TV of Japan called me from New York to ask where I would be at the beginning of October when the Nobel Prize in Physiology and Medicine would be announced. I was surprised, but thought perhaps that the Japanese media were indulging in some wishful thinking about publicity in future years, and were probably calling several senior scientists each year. I was just leaving for Belgium to help Dr. J. Mallefet with brittle star research at the time, so I gave them the name of a hotel in Brussels.

According to the *Newton* magazine in Japan, which announces the names of their candidates for the Nobel Prize each year, my name was among the possibilities, but listed at about the middle of their ranking for the physiology and medicine category; I thought it unlikely to happen and I decided to ignore it. Since that time, I had received two or three similar calls every year during the same season. It had become a sort of ritual that I think my family found mildly interesting, but none of us expected anything to come of it. We treated it as just a reminder of the season. The situation seemed similar in 2008. Nothing happened on October 6, the day of the announcement of the prize in physiology and medicine. One television company called me to say, "Sorry for this year. I hope it happens next year." Thus, I sent my reply to the University of Peking which I had been withholding, accepting their invitation to give a seminar—but I was to regret it soon. The next day, on October 7, I was pleased to hear the announcement that three Japanese physicists, Nambu, Kobayashi, and Masukawa, were going to receive the Nobel Prize. I thought the Japanese media and people must be very happy. On the same day, however, Mr. Aoki of the Japanese newspaper *Nikkei* notified me, as information acquired from Stockholm, that GFP was listed in second place for the Nobel Chemistry award. At midnight, Mr. Kawachi of the Japanese newspaper *Mainichi* called me with the same claim. However, I didn't think that information would be leaked from the Nobel committee, who keeps the names of the candidates unpublished for fifty years, so I thought it must just be someone's guess. Before I hung up, though, the reporter also asked me for permission to call again if Dr. Tsien received the prize.

I went to sleep.

However, the phone started to ring again. It was 5:15 in the morning of October 8. Akemi picked up the receiver, and said, "It is from Stockholm." It was the call from the Swedish Royal Academy on the awarding of the Nobel Prize; I was going to receive a Nobel Prize together with Martin Chalfie and Roger Tsien. I have already written about that very busy day. Without time to drink or eat, my day ended after 10:00 in the evening, with the end of the last TV broadcasting. Akemi almost lost her voice after responding to continuous phone calls, which exceeded sixty in number. I received over a hundred celebratory emails; the second one, at 5:54 a.m., was from my son Tsutomu, and I was glad because he seldom writes to me. The very first one was from Dr. Manabu Nakazono, a graduate of the Pharmacy School of Nagasaki University. My super busy day finally ended. However, it was only the first day.

9.1 Preparing for Stockholm; the Order of Culture Award

On the next day, the telephone started to ring at 8:00 in the morning, not only for celebration but also with numerous requests for interviews, lectures, and written communications. Akemi responded to them, declining the majority. Fortunately, however, there were no calls between 10:00 p.m. and 8:00 a.m. to disturb my sleep.

With the press at MBL on October 8, 2008 (credit: Beth Ready Liles).

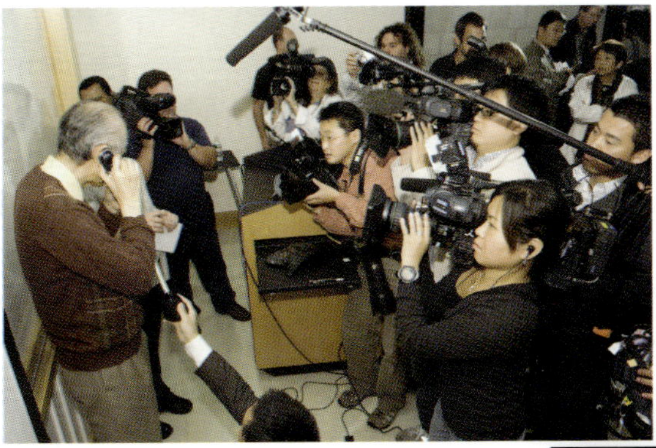

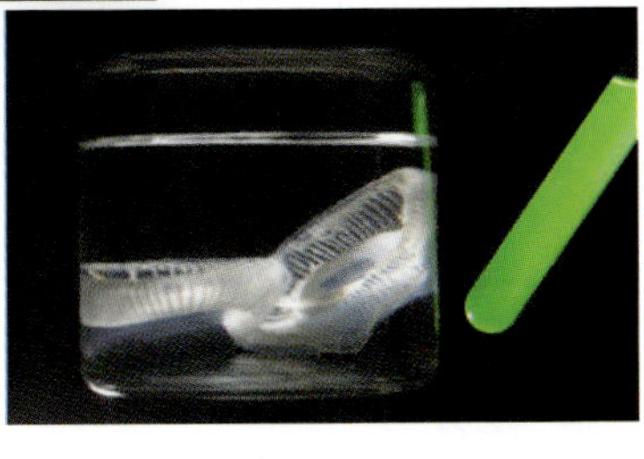

Above: press conference at MBL; lower right: *Aequorea* collected in 1999, and test tube containing GFP from 20,000 jellyfish (credit: Tom Kleindinst).

The mail package from the Nobel Foundation arrived on October 15 through MBL. According to their instructions, recipients must remain in Stockholm during Nobel Week, December 5–13, and attend several events including a Nobel Lecture (on December 8) and the award ceremony (December 10), while staying in the Grand Hotel suite reserved by the Foundation. Each recipient could invite 16 guests who would also stay at the Grand Hotel. At the ceremony and during dinner, male laureates had to wear a tailcoat, white tie, and black patent leather shoes; ladies must wear a long gown or traditional ethnic or national costume.

The choice of 16 guests was not easy. I eventually decided to invite my younger brother Sadamu Fujiyama, my sister Sumiko, and my children Tsutomu and Sachi (a total of eight persons including their spouses); Junko Ito, daughter of my late sister Setsuko; three relatives of Akemi, who came from a large family and has two sisters and four brothers living; two persons from MBL, Director Dr. Gary Borisy and trustee Professor John W. Rowe of Columbia University; and Professor Kenichiro Nakashima of Nagasaki University, and Dr. Katsunori Teranishi of Mie University. We had to submit all the paperwork by the end of October: their names, relationship, institutional position, language, hotel room preference, measurements for the rental tailcoat, and so on. At first I was not sure if my daughter and her husband could attend, because I had just found out that they were expecting a baby. Akemi and I were looking forward to the birth of our first grandchild in the spring, but we hardly had any time to think about it.

In addition to these matters that occupied Akemi and me, the Department of Science and Education (JDSE) of the Japanese government contacted me on October 10 regarding the award of the Order of Culture. I was told that all recipients of the Nobel Prize receive it unless one refuses. Usually it was given on November 3 directly by Emperor Akihito himself. However, considering my condition of poor health at that time, I requested that I receive it at my home, even though I regretted missing the honor of meeting the emperor. But I did not think I could go to Stockholm in December if I put more strain on my energy and health traveling to Japan the month before.

However, after some discussion, the JDSE asked us to hold a meeting to address the improvement of basic scholarly education after an award ceremony at MBL. I requested Dr. Shinya Inoue of MBL and Dr. Susumu Honjo of Woods Hole Oceanographic Institution (WHOI) to join this meeting, and arranged the meeting place at MBL. I also put in a request for use of the desk and chair that had been used by the late Emperor Showa (known during his life as Emperor Hirohito, he was also a scientist, a marine biologist) when he visited MBL in 1975.

On November 8, Mr. Isoda and Mr. Goto of JDSE, and the Consul General in Boston, Mr. Suzuki, and Consul Kobayashi arrived at MBL, and the delivery ceremony of the Order of Culture was carried out at the Grass Reading Room of the library. Pamela Clapp Hinkle, Diana Kenney (science writer at MBL), and two or three other people were there to represent MBL. Pamela was the chief of the communication office at that time, and I knew her well since she was the editor of the *Biological Bulletin* published by MBL.

Since MBL director Gary Borisy was out of town, I asked Pamela to represent MBL. In the delivery ceremony, people suggested that I sit on the chair that was used by the Emperor Showa. I hesitated to do so, but sat on it to have a picture taken. Consul Suzuki hung the Order of Culture medal around my neck. It was fashioned in a mandarin-orange flower design, with five petals; the mandarin tree symbolizes eternity. It is a great honor, but I wished I had received it before the Nobel Prize announcement. We toasted the occasion with the champagne that Consul Suzuki had brought. Then we discussed the strengthening of fundamental scholarship in Japan.

Although I was not able to receive my Order of Culture award directly from Emperor Akihito, a pleasant surprise occurred that allowed me to meet the emperor after all. When I returned to Japan the following March, for the first time since the Nobel ceremony, the Emperor and Empress of Japan invited Akemi and me to a dinner at their residence. It was not a gorgeous dinner party at the palace, but I was very much honored by this rather simple dinner for only the four of us in their private quarters in the "Gosho," as the residence of the imperial family is generally called. The meal was in the traditional

Japanese style with soup, tempura, and other unpretentious dishes. But the ingredients of the tempura had been specially picked by their majesties in their afternoon stroll in the palace garden. I found that they were interested in the wild vegetables growing ripe in each season in the palace. The Japanese-style soup was especially excellent; I had never tasted such a delicious soup.

I think the emperor gave me a special meeting to compensate for my absence at the formal occasion on November 3. Emperor Akihito is a scientist like his father, so we talked much about science, especially about his recent studies. We also talked about a lot of the hardships caused by World War II because the emperor and empress were of the same generation as Akemi and I, and had had plenty of sad experiences from the war. The Japanese emperor and empress are such a warm-hearted considerate couple. I am very proud of this as a Japanese citizen.

With the Order of Culture medal and my book on bioluminescence, in the MBL Library, November 8, 2008 (photo by Tom Kleindinst).

In the middle of October, not long after the Nobel announcement, Pamela asked me to attend a dinner of MBL trustees with Akemi. I wondered about its purpose. After the dinner, they gave me a framed appointment paper indicating that I was now a Distinguished Scientist, the highest research position at MBL. I was the third person to receive that position. It seemed really impractical to me, however, from a scientist's perspective, to appoint a retired person such as myself to a position that should be intended for a currently active researcher.

However, the action of Boston University seemed even more illogical. Although I had retired from Boston University in 2001 when I retired from MBL, three days after the announcement of the Nobel Prize, on October 10, Karen Altman, Dean of the Medical School of Boston University, sent me an email saying, "Because your title has not been changed yet, we will change it now. You are now honorary professor of physiology and biophysics." Thus, it meant that I was suddenly an honorary professor since several years ago! My retirement, which had happened without anyone seeming to take much notice at the time, was now the occasion for this honorary position. Perhaps, though, it was an appropriate continuation for my path. After all, I had acquired my first PhD in Japan without embarking deliberately on the degree program, but rather by being offered it incidentally as a result of my research and the Fulbright Fellowship. So my entire career was, in that way, illogical.

I continued to decline all interview requests, except one from *Newton* magazine of Japan. I also spent a couple of days each to respond to Swedish Nobel TV and STV (Swedish Television), requests which I thought that I had a duty to accept. On the other hand, I knew I had to set aside adequate time to prepare for my Nobel Lecture, which would probably be the most remembered and publicized lecture of my life.

The Nobel Foundation provides laureates with an attaché who takes care of paperwork and facilitates various matters during Nobel Week. My attaché, Marie Andersson de Frutos, called me on November 17 for the first time, and then called me every three days until my departure to consult on various questions. She normally works at the foreign ministry department. Her involvement was indispensable. I knew I could count on her to help me solve all etiquette and practical issues relating to Nobel Week, matters that I found much more esoteric and confusing than lab work and bioluminescence research.

In addition, I received invitations to dinner from the Swedish ambassador in Washington, DC and an invitation to visit the White House. However, I was in poor health due to the continuous stress since October 8, so I had to decline both offers. Anyway, I did not care to shake hands with President Bush. I thought later that if President Obama had already been in office, I would have been very interested in meeting him. I suffered from a sore throat and digestive problems during this time. Twice I visited the emergency room of Falmouth Hospital. My doctor, Robert McGowen, told me not to go to Stockholm, and that if I must go, to limit the number of events to two per day.

I told this to Marie. Perhaps there was a talk between my doctor and the Foundation. I was told that I must attend the ceremony and the dinner in addition to the lecture, but that I could skip all other matters if I could not attend. My invited guests included two medical doctors, my younger brother Sadamu Fujiyama and my oldest nephew, Yutaka Yamane (the son of Akemi's oldest sister), and they would be able to take care of me in case of an emergency.

As the departure date approached, we had many calls from the media asking about my itinerary, dates and route of travel, airlines, who would accompany me, and so forth, but we did not give them the date and the airline. I knew they would find all that out anyhow, in some way or another.

9.2 To Stockholm

In the evening of December 4, my wife and I took a reserved limousine to Logan Airport in Boston, and then boarded an Icelandair flight to Stockholm, via Iceland. The airport there is fairly small, calm, clean, and very pleasant. We have never found it too crowded, and we knew from previous travels that the food served by the airline would be good. Our flight took about eight hours, arriving at Stockholm about noon. I noticed that the older brother of Dr. Tsien, who is an engineer and who had greeted us in Iceland while we were waiting for the connecting flight, was on the same plane. When we got out of the plane at Stockholm, two persons met us: Professor Gunnar Öquist of the Swedish Royal Academy and my attaché, Marie. Led by her, we went down the steps to the ground. A limousine was waiting for us, and we were then taken to a VIP room with a red carpet. The VIP room had a cherry motif, and a line of cherry trees was planted in the middle of the corridor, although their flowers were artificial in December. While we were talking with Professor Öquist, our three pieces of luggage were brought in. We went through no customs or immigration. We really felt like we were national guests. Another laureate arrived from a European country while we were leaving the VIP room. Then we were driven to the city of Stockholm in the eight-passenger limo (I think it was a Volvo), which was provided for our use during our stay for Nobel Week. Therefore, the people waiting for us at the airport never saw us.

Marie called the hotel and found that many reporters were at the hotel entrance. She asked the Grand Hotel to open a side entrance, then we proceeded directly to our room on the fifth floor. Marie took every effort to care for my poor health and especially paid every attention to keeping us from being bothered by the media; she was wonderfully resourceful in that respect, and indeed, marvelously adept at organizing all the details that could facilitate a smooth Nobel Week for us. Our room had a spacious bedroom and a study, and was decorated with flowers. I also found fruits and candies. One of the flowers was a gift from Mrs. Teranishi of Japan.

Two to three hours later, my brother Sadamu and his wife arrived from Japan, and my son Tsutomu arrived from California with his partner Julia Menapace. We wanted to have dinner together, but all the restaurants in the hotel were full with laureates and their families, and the crowds were also large because it was Friday evening. Then, Tsutomu found the hotel manager, who prepared a special room for the six of us, and we were able to have dinner. Unfortunately, we older people were already very tired from the long trip, so that only the younger members of our group could really enjoy that first dinner much.

9.3 The Nobel Week Begins

On December 6, 2008, the next morning after we arrived in Stockholm, I overslept because of jet lag and the darkness due to the high latitude, so I decided to skip breakfast. At 9:30, Marie came to pick us up. We left the hotel for a "get-together" to meet other Nobel prize recipients at the Nobel Museum. Marie said that many reporters were at the front entrance of the hotel waiting for the laureates to appear, so we used the same side exit we had used the previous day to avoid them. In the light rain, many reporters waited for the arriving laureates by the doorsteps to the museum. Most of them were Japanese. Marie opened the car door, and as soon as I stepped out of the car they came up close to me, but Marie and the driver, Mikael Sjoblom, shielded me from them, and I entered the Museum without being bothered by the reporters.

This was the first meeting of Nobel Week, and the Foundation staff warned us all about the coming strenuous week. Yes, they were right, and it turned out to be just so, especially for an older person like me. The museum focuses mainly on Alfred Nobel and the Nobel Prize. There was a rail attached to the ceiling from which the photographs of past laureates hung and were slowly rotating. There was a special event on the "Culture of Creativity" going on. A staff member asked me to sign under the back of a chair with a marker pen. The chair would be used in the museum cafe; therefore the guests would drink and eat sitting on the chairs signed by laureates. After the events for the start of Nobel Week, we had a buffet lunch there. We had realized that Swedish people eat reindeer commonly; I had never before tried it.

I was also interested in Nobel ice cream, but I was unable to find it. In the afternoon, an official portrait photograph was taken at the hotel. I then cancelled a personal interview for the Nobel website. I had already spent enough energy by that time, and the week was just beginning.

On December 7, I requested a wake-up service. It was a very long day, and I was very tired from the previous day. At a breakfast meeting at 8:00 a.m., with other laureates and chief members of the Royal Swedish Academy of Sciences, I met with Professor Gunnar von Heijne, Chief of the Chemistry Department, for the first time. The Academy of Sciences was housed in an old historical building. A press conference occurred at 9:00 a.m. after breakfast, and then we visited the Aula Magna building of Stockholm University, where the Nobel lectures would be given. We heard that the Nobel Prize was determined and the telephone calls were made from there on October 8, though perhaps the calls were from the Academy of Science building instead. So many things happened that week that I cannot remember for sure. We had no time to take notes.

In the Nobel lecture rehearsal, I made sure that my PowerPoint worked correctly and that my GFP sample would show enough light to be seen at the top of the hall. Then there was a news conference at Stockholm University. I think the person who received the most questions from reporters was the winner of the economics prize, Professor Paul Krugman.

In the afternoon, we visited the Archive of the Royal Academy of Sciences. I saw a book by Copernicus (1473–1543) and other old books. Particularly noteworthy to me was the writing of Carl Peter Thunberg, a Swedish botanist, who stayed on Dejima in Nagasaki (an island made in the bay nearly four centuries ago to serve as a European trading post) during the Edo period when Japan was closed to other countries except for Dutch ships. His original record of the "Speech on the Japanese Nation 1784," his lecture made as the president of Uppsala University, was preserved there, and we three Japanese laureates were presented with its English version published by the Royal Swedish Academy of Sciences in celebration of the famous botanist Carl Linnaeus's 300-year birthday in 2007.

Then we attended the reception by Professor Gunnar Öquist at the Royal Swedish Academy of Sciences. It was very crowded because all of our guests were also invited. Next we attended a dinner party for laureates and their spouses at The Villa, in an old room with the portraits of the presidents of the Academy. After the dinner, a dessert was served while we listened to a violin quartet in the other room. In the afternoon of December 7, my daughter Sachi and her husband John Brinegar arrived from Richmond, Virginia; my sister Sumiko and her husband Fumihide Sawayama, and my nieces Junko Ito and Kaori Okubo also arrived. I think some of them gathered together for a more casual dinner, or met up in the evening together with Tsutomu while Akemi and I were busy attending official events.

On December 8, I had breakfast for the first time at our hotel. Since the famous Viking breakfast room was very crowded, we quietly had a continental breakfast in the tea room. I saw photographers aiming at us from the street outside. It was now the day of my Nobel Lecture. My lecture time was at 12:35, so I prepared for it in the morning and headed to the Aula Magna at Stockholm University at 11:30. My lecture time was thirty minutes. After an introduction by Professor Gunnar von Heijne, I gave my lecture: "Discovery of the Green Fluorescent Protein GFP." The 1,200 seats of the hall were fully occupied. In this way, I finished one of my obligations. I took my lunch with Marie on the seventh floor, at the top of the hall, then, together with the two other chemistry laureates, had an interview with reporters.

There was going to be a reception at the Stockholm Concert Hall. But I had a bit of a cold then, as well as other ailments aggravated by the stressful previous two months. Considering my health condition and the advice my doctor had given me, I skipped attending it and the Nobel Prize Concert following the reception, because that was expected to continue until late into the night. However, Akemi attended and enjoyed the music. Since I had only eight tickets for my sixteen guests, I gave my own ticket to Tsutomu so that he could escort his mother, and Tsutomu gave his ticket to his partner Julia; I was later told that their seats were close to the seats of the king and queen.

Later that night, the Yamane family arrived—my nephew Yutaka with his wife and two children. He was able to advise me on some medicine for my cold. Akemi's younger brother Akihiro had also arrived. He provided some comic relief one evening in our suite by asking to put on my Order of Culture medal, then posing comically for some pictures. Tsutomu and Sachi were there at the time, and also other relatives; we all laughed.

October 9 was not a particularly interesting day for me. I skipped attending the TV discussion at the Royal Library at the Palace in the morning. Both Tsutomu and Sachi went there with the Nobel invitees, and enjoyed a tour of the palace, though Sachi was often tired because of her pregnancy. Marie accompanied them. We all attended the buffet lunch reception given by the Ambassador of Japan, but I had no time to eat anything there due to continuously meeting Japanese residents. Tsutomu and Sachi appreciated the chance to talk with some of their aunts, uncles, and cousins, whom they had rarely been able to meet up with over the years. We had only traveled to Japan as a family once or twice every decade. Some Japanese relatives had visited us in Princeton or Falmouth, or attended Sachi and John's wedding several years ago, but visits had been spaced years apart because of the costs and time needed for the fourteen-hour flight. Then, following the lunch reception, I had an interview with Japanese media. Later, a reception was held by the Nobel Foundation for all attendees of Nobel Week at the Nordic Museum, which I skipped, but Akemi attended. She liked viewing the exhibits there.

Because of the warning that restaurants in the hotel would be closed for the preparation of the Nobel banquet next day, I asked Marie to reserve us a restaurant in town. It was a small but extremely nice restaurant. We were all satisfied, but unfortunately I have forgotten the name of that place.

9.4 The Nobel Prize Ceremony and the Nobel Banquet

The next day, December 10, was the day of the award ceremony at the Stockholm Concert Hall. We left our hotel at 10:00 a.m., and the rehearsal began when all the laureates had arrived. Professor Kobayashi was to come onto the stage first, followed by Professor Masukawa, myself, Dr. Chalfie, and Dr. Tsien, led by a young lady in a white dress with yellow shoulder ribbons. When the music started, we walked onto the stage and took our seats. Then we received instructions on how to behave when the king awarded our prizes. We were told that we must remain standing up while the king was standing. Then the rehearsal ended.

After having our lunch from room service, we put on the tailcoat and formal gown; it took me some time to fasten the cuffs of the shirt. Tsutomu and Sachi had taken the opportunity to see the old town near the hotel during our rehearsal, and had indulged in some casual shopping and lunch. Afterward, Sachi came to help her mother dress. I heard later that John helped Tsutomu with his cuffs; John had had prior practice with his wedding finery. (Nobel Week requires much more formal wear than scientists or academics usually need.) I think both their aunts wore formal kimonos, but Akemi decided to wear western dress. I wondered if I should wear the Order of Culture medal or not, but in the end I decided to wear it.

At 4:00 p.m., we returned to the Concert Hall. At 4:30, the sound of the royal anthem "Kungssången" began, and then we entered onto the stage from a side entrance, as we did in the rehearsal. On the stage there were the king, queen, prince, two princesses, officers of the Nobel Foundation and Royal Academy, and also about one thousand people in the viewers' seats. Many of my invited guests were seated in the first two rows. After the king took his seat, we all sat down. The chairman of the Board of the Nobel Foundation, Dr. Marcus Storch, talked about the purpose and objectives of the Nobel Foundation, followed by the chief of the physics department, and then the physics prizes were handed to Dr. Kobayashi and Dr. Masukawa by the king. (Another physics laureate, Dr Yoichiro Nambu, was too old to attend the ceremony and I heard that he received his prize at the University of Chicago.) Then, after the explanation by the chief of the chemistry department, Professor Måns Ehrenberg, the king handed the chemistry prizes to me, Chalfie and Tsien. It was interesting that a simple explanation and congratulation was added in Japanese in the case of Japanese awardees. At the end of the ceremony, the king, queen and princes left while we stood and music played. On the stage, we shook hands with the officials of the Nobel Foundation and members of the Royal Academy of Science, as well as family members who came up onto the stage. All through the ceremony the music was performed by the Royal Stockholm Philharmonic Orchestra.

To attend the celebration dinner, the magnificent Nobel Banquet, we moved to the City Hall in light rain. It was only a ten-minute walk, but felt very cold; we were fortunately able to use a limo. The City Hall was an old historical building and could provide dining space for 1200 people. We had to wait for a long time on the second floor while all the general guests were being seated in the huge dining room, the "Blue Room," on the first floor. We formed a line in the sequence of our given numbers. Akemi was No. 4, and her partner was the speaker of the Swedish Parliament, Mr. Per Westerberg; I was No. 5 with Mrs. Storch as partner. No.1 was King Carl Gustaf and Professor Françoise Barre-Sinoussi, the recipient of the physiology prize, No. 2 was Queen Silvia and Mr. Storch, No. 3 was Crown Princess Victoria and Prof. Chalfie.

Then the music started. I took the arm of my partner and went down the staircase, proceeding to the main table at the center of the hall. All of the common guests were standing up. When Mrs. Storch showed me my seat, I stopped. When everybody was at their seat, the king took his seat first, and then everybody else was seated.

First, led by Dr. Storch, we toasted King Carl Gustaf, then the king toasted Alfred Nobel, and then the dinner began. The menu was sole and shellfish, followed by filet of veal with vegetables in a superb presentation. Several hundred servers with dishes came down the stairway and waited at the back of each guest. The king's dish was served first, then the dishes of all the others were served at the same time. There were some opera-like attractions. On my right side were Mrs. Storch and the king; the queen was across the table from me. On my left side was Dr. Hazelrigg, the wife of Prof. Chalfie. In a conversation with Mrs. Storch, she said she had been in Japan only once and did not travel much, although her husband traveled frequently, especially to New York. Akemi's seat was across from the king, on her right side was Dr. Storch, the Chairman of the Nobel Foundation, and on her left was Mr. Per Westerberg, the speaker of the Swedish parliament; it was a place of high honor. Mr. Westerberg was a person skilled in conversation. He talked to my wife about the old history of the city of Uppsala, and the prosperity of Sweden at the time of building this city hall. The conversation between Akemi and Mr. Storch began when he found out that our marriage was an arranged one. He showed interest and asked if my wife and I had seen each other before marriage. His parents also had had an arranged marriage; their marriage was decided without their seeing each other but continued for forty-seven years. He concluded that therefore arranged marriage seemed not to be a bad idea after all.

At the end of the dinner, there were three-minute speeches by representatives of the prize recipients in physics, chemistry, physiology and medicine, literature, and economics. Dr. Tsien made a speech as a representative of the chemistry winners; he is a lively speaker, easy to listen to, and I was most happy that I did not have to give the speech on this occasion. He mentioned in his speech that the jellyfish *Aequorea*'s decline was not caused by our lab's overfishing because other kinds of jellyfish have also disappeared from Friday Harbor. The great dinner party ended with these speeches. All of the main table guests followed the king and queen upstairs with their partners while music played.

Thereafter, we waited for a long time on the second floor. Marie asked me to give an interview to Swedish TV for one minute, but there was noisy dancing going on and I could not hear well. When

I tried to return, I was blocked by a police line; beyond it I saw that Dr. Kobayashi and Dr. Masukawa were talking with the king and his family. When it was the turn of the chemist laureates, I had a very short time to talk with the king because we were six people. Because the king looked at my Order of Culture medal, I said, "I received this when it was announced that I would receive my Nobel Prize and this is the only one I have." The king replied, "I also have only this one," indicating the medal he was wearing. The crown princess was highly interested in environmental problems; therefore I talked with her about the contamination at Friday Harbor. We were unable to leave the hall until all meetings with the king were finished, probably for security reasons. The Nobel medals and diplomas of all laureates were displayed in a case in this room.

We returned to the hotel after 11:00 p.m. Later I learned that Tsutomu stayed at the dance and the other events until morning. He mentioned that some of the laureates had remained late, too—I think only the youngest ones and the students must have enjoyed the party all night! It was a long day and I was tired, but I was glad that I had finished my duty.

9.5 The End of Nobel Week

On December 11, I cancelled a sightseeing tour of Stockholm in order to attend the lunch party given by the Education Minister of Japan, Mr. Shioya. I was a little sorry, because this meant that I was unable to have any tour of Stockholm. The lunch party was at a Chinese restaurant in the city. I was told in advance that it was a rather important meeting with the minister, but I seemed to have misunderstood; there was no such talk.

After returning to the hotel, I did not feel well and cancelled my dinner with the king and queen at the royal palace. Next morning, the Ambassador of Japan kindly telephoned me to check on my health.

On December 12, the last day of Nobel Week, I went to the Nobel Foundation, signed for the receipt of the prize money, and received my diploma and Nobel medal. All of the laureates were photographed together. I invited Marie, Tsutomu, Julia, Sachi, John, and my driver Mikael to dinner at a fish restaurant.

At 7:30 a.m., December 13, someone knocked on the door. I opened the door in pajamas and found five young people in white gowns, together with a hotel boy with morning tea. They came in and lined up at the end of the bedroom. In addition to the white gowns, the three women in the middle wore red belts and the boys at both sides wore white conical caps. They had lighted candles in their hands, and the woman at the center had five lighted candles on her head. They sang some carols and moved to the next room. I learned later that the day was Saint Lucia's Day, a Swedish national holiday. Lucia was a Christian woman in the second century who died refusing a marriage with a man of different religion, and who was given the title of saint later. She is one of the seven female saints commemorated by name in the Roman Catholic Canon of the Mass. On the morning tea tray, we found a small gift from the hotel: a clay candle stand in the shape of Saint Lucia. Later Lucia's carol members came down to the hotel restaurant and went all around. We enjoyed their sweet voices again during our breakfast.

We left at 11:00 in the morning from the hotel on Mikael's limo, heading to Arlanda airport outside Stockholm. At the hotel entrance lobby, Marie's husband and their two children, both boys, were waiting to greet me; they had taken the trouble to drive in from the suburbs of Stockholm that morning. It was a very nice occasion for me. Marie later sent me her boys' beautiful picture as she had promised, and Akemi and I treasure it together with our memory of her and the special week we spent with her. The boys must have grown up into fine young men by now.

Marie and Mikael checked our baggage for us. We waited in the same VIP room where we had waited upon our arrival. Our plane was delayed a little. Shortly before our departure, Dr. Kobayashi's group also arrived. Mikael took us to the door of the plane in his car, and we said goodbye to Marie and Mikael who had taken such wonderful care of us for one week. We arrived in Boston via Iceland and caught our limo, arriving at our home at 8:00 in the evening.

Later we learned that Tsutomu attended the Lucia Ball of Stockholm University on the evening of December 13 in my place. The Lucia Ball is traditionally the end of Nobel Week; it is a pleasant meeting, though a little foolish, at the end of the serious events of Nobel Week. Although I don't yet know the reason, the main subject is a green frog, and attending Nobel laureates jump like frogs. But no media are allowed there. I later saw a photograph showing Tsutomu, Chalfie, Tsien and Douglas Prasher carrying a large frog. Tsutomu told me that he received a green frog pendant intended for me on that occasion.

CHAPTER 10

After Stockholm: Visiting Los Alamos, Remembering Nagasaki

Although Nobel Week was over, I still had one further immediate obligation. Each Nobel laureate must submit his Nobel lecture and autobiography to the Nobel Foundation for their yearbook by the end of February in the year following the award. Thus after returning home from Stockholm, I tried to concentrate on writing my autobiography. My time, however, was much interrupted by other work, such as responding to requests by the media, even though I tried to keep my involvement to a minimum, as well as making all the essential preparations for my planned trip to Japan. Many events were arranged at science societies, universities, and at the schools where I graduated, and so on, so that the necessary details, correspondence, and scheduling kept me extremely busy. Because I was not used to writing an autobiography, I asked help from Diana Kenney, a science writer at MBL, in producing a rough copy. She was very helpful.

I still wished I had more time to work on polishing it. Perhaps writing the autobiography rapidly is a little easier for the younger laureates, whose lives and research careers had spanned less time. It had been sixty years since I had begun studying chemistry, 7,000 miles away in Isahaya, in post-war Japan. Over 45 years and 3,000 miles separated me from the beginnings of my research that would, through twists and turns, lead to the discovery and development of GFP, and set the foundation for my Nobel Prize. The world had changed much in that time. Japan, the United States, and even the sciences had changed unimaginably since my birth.

10.1 My Welcome in Japan

Although I have now lived in the United States for the majority of my adulthood, and always write my scientific research results in English, I am still Japanese at heart. So I felt it was my responsibility to support the progress of science in Japan with whatever visibility my awards had given me. In March of 2009, in Tokyo, the Chemical Society, Pharmaceutical Society, and Agricultural Chemical Society of Japan jointly organized my lecture. After that, both Nagoya University and Nagasaki University requested that I give public lectures. The trip to Japan was welcome, but extremely busy.

At the Nagoya campus, I and two other Nobel laureates of that year who were alumni of the university each planted a tree in memory of this occasion. The tree was a species called "ranshinboku" in Japanese, but I do not know further what kind of tree; it is not a very common name, I think. Through its growth, it would help commemorate the occasion. I thought a tree was better than a stone monument, and wished other places also would use that idea. Also at Nagoya, I was very pleased to receive a gift of a beautiful cloisonné vase from the governor of Aichi prefecture. Aichi prefecture is the region within which the city of Nagoya is located and of which it is the capital. This region has been known for its cloisonné enamel since the late 1800s; in the Meiji-era industrialization of Japan, German techniques were brought in to further develop Japanese cloisonné techniques. The art work on the vase was specially made for me with a graceful stylized jellyfish design on it. In addition, as a special feature, the jellyfish on the vase glow blue-greenish under ultraviolet rays. I received many splendid gifts with sincere thanks, but this was one of my favorites.

At Nagoya I particularly enjoyed one pleasant and convivial occasion. The old members of the Hirata lab, almost twenty persons altogether, gathered at dinner and talked about old times, especially the hard work of *Cypridina* collection that each had experienced more than half a century ago at the Inland Sea. Interestingly, in these days in the town of Setoda in the Inland Sea, some volunteers, inspired by our past collections, have opened a summer camp for school children to learn about *Cypridina*.

While at Nagoya I was able to visit the cemetery of Professor Hirata with his widow and two daughters. When I put my head low in front of his tombstone, I felt as if I heard him speaking in his usual gentle voice with a smile, saying to me, "You did well."

At Nagasaki, the Governor and Nagasaki University held a joint celebration party for citizens (representatives), friends, alumni, and relatives. The elementary school brass band members played wonderful music there, I remember. The place was crowded with excited people, so that I was not able to have a chance to speak even one word to everyone there. I regretted that very much. Among the attendees were many friends who had come from far away, to see me again after 50 years.

At Sasebo, the Sasebo Minami High School desired to install a memorial stone commemorating my Nobel Prize in the corner of their yard, to encourage future students. I had once attended this school, before the war, when it was still a middle school. After the war my old middle school campus changed into a high school. At the same time, the old system, in which students attended middle school for five years, was replaced by a new system in which middle school and high school each became three years long. The city of Sasebo also established a science contest with my name for their public school children. Since 2009, about 3,000 children in grades 1–9 work on their own projects each summer, and the best students are honored in the fall.

Returning to Tokyo, I then visited the Prime Minister with whom I had talked on the phone on October 8. Photographs were snapped, and then I was taken a short distance by car to greet committee members of congress. I also received a kind invitation from the Swedish ambassador in Tokyo at that time, but my tight schedule did not allow me to accept it. I still feel regretful at having to refuse.

The last event was a visit to the emperor and empress at their residence. This was a building separate from their palace where all public affairs are held. We received the invitation originally by an inquiry from the Imperial Household Agency and arranged the date and time first by email and later by phone with the person in charge in Tokyo. I have already talked about that visit in an earlier chapter, and how impressed and honored we were at their gracious and unpretentious hospitality. When leaving the palace, we noticed the cherry trees were in bloom by the moat; we could see water glinting under the cherry blossoms. Here, at Edo Castle, built for the shogunate in the Edo period and surrounded with a moat, they showed their soft color in the evening light. This quiet scene gave me a welcome peace of mind after my busy schedule for the last two weeks, or, indeed, for the past several months after the announcement of the Nobel Prize. I felt for the first time that maybe receiving the prize was not such a bad thing after all.

10.2 Lindau, Germany

As a Nobel Prize recipient I received many requests to attend meetings. Letters poured in from Japan and from all over the world during the following years. Those invitations were not only to science conferences but also included social or world affairs meetings among them. One that was particularly striking was a very elaborate invitation card from the king of Morocco to attend a conference to be held in his country. Among these invitations, I could only accept a few, but we attended the 59th and 60th Meeting of Nobel Laureates at Lindau in Germany in the summers of 2009 and 2010.

In the picturesque island city of Lindau located at the eastern end of Lake Constance, students and researchers from all over the world meet for a week in the summer. The atmosphere is invigorating. I found that the young researchers, particularly Asian researchers, were very eager to learn from us laureates and I wish I could have participated more often in this event, but my poor health, especially in the summer time, prevented me from returning to the meeting thereafter. There were many elderly laureates in attendance and they do their best to encourage the younger scholars. Under the guidance of the president of the Council for the Lindau Meetings, Countess Bettina Bernadotte, I think this meeting is really working wonderfully for both young researchers and laureates.

In several Asian countries, not only are young scholars encouraged to attend, but some government agencies also support this meeting to inspire their own country's students. Princess Maha Chakri Sirindhorn of Thailand's royal family, a member of the Honorary Senate of the Foundation for the Lindau Meetings, attends this meeting every year as a science adviser of her country. When I attended one of the science conferences held in Bangkok, Thailand, I had the honor to sit next to the princess at the banquet. At Lindau our hotel room happened to be the one next to the Princess's suite, and the corridor in front of her room and leading up to our room was overwhelmed by the many flower bouquets presented to her by German people.

It reminded me of when I had visited Washington, DC in 1962 on the occasion of the Fulbright meeting, and my wife and I had had to wait for a while to enter the exhibition room at the National Gallery of Art because Thailand's queen was visiting there. A short time later she appeared at the entrance to leave, so we happened to see her close up, and were very much impressed by the pretty queen wearing a yellow dress. In my father's photo album there was a picture of the young King of Thailand. During the Second World War, my father spent time in Thailand as a military officer and had to remain there after the war to negotiate the Japanese soldiers' safe return to Japan. The pretty queen whom we glimpsed must have been the King of Thailand's wife and the mother of Princess Maha Chakri Sirindhorn.

Also, when I received an honorary degree from Gakushuin University of Japan as the seventh such person, the Princess happened to be the sixth person to receive that same honor. Thus I felt some strong connection to the royal family of Thailand. Princess Sirindhorn is a very energetic person, and she uses her energy productively not only in attending science conferences, but also on the occasion when a flood disaster happened in her country; she was extremely active in providing care to the victims. She is a very popular figure among the royal family, an admirable supporter of younger scholars, and someone whom I was very pleased to meet again.

10.3 Research in Russia

My only direct participation in research after the Nobel award was in Russia. By attending many meetings I had lost my own research time, regrettably. When I had visited Dalian (known as Dairen in Japanese) in China at one such meeting, Dr. Josef Gitelson, a member of the Russian Academy of Science, visited me and asked for my collaboration with his group's research on bioluminescence. Overcome by his eagerness, I agreed to work with him to obtain a mega grant from the Russian government.

As a result of the successful grant application, I was committed to visiting Siberian Federal University and the Institute of Biophysics in Krasnoyarsk, to conduct discussions with the Russian researchers. Krasnoyarsk is located by the Yenisey river in Central Siberia, and is situated in the middle of Russia geographically. We traveled there three times, in 2012 and 2013, after obtaining our Russian visas in Tokyo during an earlier trip to Japan in the spring of 2012. To try to make our many travels more efficient, we combined some of our travels to China or Russia with flights to Japan. Still, we spent much time in transit, and Akemi and I needed to put much thought and energy into the logistics of our travels.

Some of these travels brought back powerful memories of the past, as my present path intersected with places or events that held strong remembrances from many years ago. Dalian, then part of Manchuria, had been a part of my childhood memories. I had last visited there with my Aunt Matsue more than seventy years earlier. This was one of the first of many journeys that I would take. At that time, as a Japanese boy in Manchuria, I never imagined how far I would go over the years, or how much the world around me would change during that era. Another memory these travels recalled was of my beginnings as a true scientist. When we accepted an invitation to Nagoya in 2009, I saw for the first time the PhD diploma that Dr. Hirata had obtained for me in 1960. Since he had done all the paperwork himself, while I had been busy preparing to leave for Princeton, I had never seen it before. A third major remembrance involved my father, who passed away during my stay in Princeton. In May of 2012, we took Sachi and her family to Japan, and we held the fiftieth year memorial service for my father in a Buddhist temple. (It is a Buddhist custom to hold memorial services at 1, 3, 7, 13, 17, 23, 27, 33, and 50 years after the person's passing.) In the memorial service, while the priest chants a sutra, participants, who usually include all the closest living relatives and representatives from the families of their extended relatives, go up in turn of their relationship to the deceased to burn incense and pray for the peace of the departed spirit. Tsutomu also joined us in Japan for this event; as a direct male descendant, his presence was expected.

Port Arthur (Ryojun in Japanese), Lüshunkou district, Dalian, China, 2011.

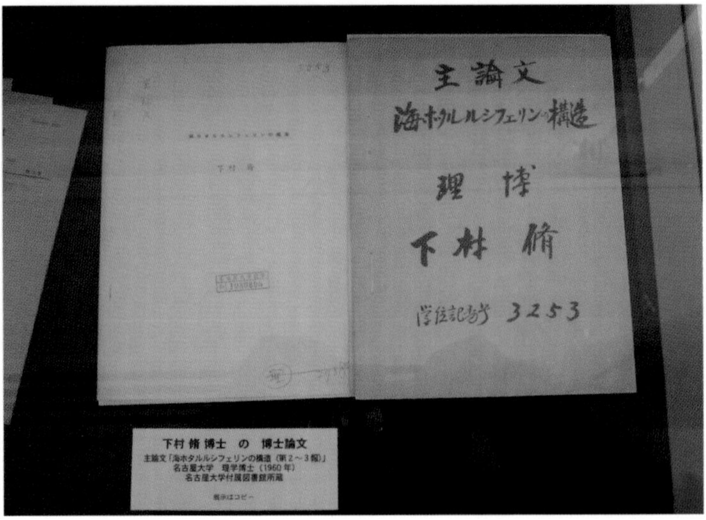

My PhD diploma on display, Nagoya University.

After Sachi and Tsutomu left Japan, we flew to Russia with a change of planes in China, and were welcomed by Dr. Gitelson's group. It was an unusually warm spell for them. We found Krasnoyarsk to be a large city set along the Yenisei River, with a population over a million, and many scenic areas and parks set near the river. We thought our granddaughter would have enjoyed running around in them. The scientists took us sightseeing between talks and research obligations.

I think under Dr. Gitelson's supervision the researchers of his group used their federal research grant effectively and made good progress in bioluminescence study. One area of their work that I thought particularly interesting was their research on the earthworm that lives in the soil of Siberian forests. It gives off weak luminescence and each specimen is a thin, thread-like creature, just 1 millimeter wide by 20 millimeters long in size. To extract its luciferin, they had to collect a lot of specimens every autumn from under the leaves. This task reminded me of our laborious collection of jellyfish, but it must have been much harder than that because of the minute size of the worms. The scientists who collected them also depended on help from family members, like we had done. In bioluminescence study the greatest difficulty is always in being able to collect enough specimens to extract enough pure luminescent material—many years of work may go into obtaining a scant few milligrams of luciferin or photoprotein. However, no scientific progress could occur without this laborious and time-consuming work to produce enough purified substance that may then allow researchers to detect chemical characteristics beyond our current knowledge.

10.4 An Invitation to Los Alamos, Where the Atomic Bomb was Built

One other visit left a particularly lasting impression on me, because it tied back so closely to my own experiences, and also my wife's. Since the aftermath of the atomic bomb in 1945, when I saw such suffering and shocking scenes of harm and death, I have tried not to recall memories about the impact of the atomic bomb, but it was always at the back of my mind; my awareness of it has never vanished.

In 1995, on the fiftieth year memorial of the atomic bomb, my local newspaper, the *Cape Cod Times*, asked me to contribute my recollections for an article. I thought it my duty to help younger generations understand what that experience had been like, and how indelible its impact was on those who had lived through it. My story was printed side by side with the one contributed by the captain of the B–29 who dropped the bomb upon Nagasaki. He also lived on Cape Cod before he passed away several years ago.

Then, in February 2012, I received a letter of invitation from Dr. John Pearson of the Los Alamos National Laboratory, who I think is a friend of my son Tsutomu, asking me to give a lecture there. The laboratory is well known as the place where the atomic bombs (Little Boy and Fat Man) that killed a great many people in Hiroshima and Nagasaki were designed and made through the secret Manhattan Project. I was very interested, but at the time my schedule was full. Then, in November of the same year, I received an official invitation from the director of the lab, Dr. Charles McMillan. I thought that I could not refuse or postpone this trip, which would take me close to the origins of the bomb that had loomed so large in my life. So I arranged to give a talk on April 18, 2013.

This visit perhaps held even stronger meaning for my wife. Like me, Akemi had also felt a very strong sense of her destiny being connected to the atomic bomb through the events in her life. She was only nine years old when her older brother was killed by the blast. She had been much closer to ground zero than I was, actually within Nagasaki on that fateful day. She was inside a house at the crucial time, so that she did not actually see the planes delivering the bomb. Had she been outside, she might not have survived. Certainly she would have been exposed to more radiation. Like me, after the war, Akemi had gone to the rebuilt Nagasaki Pharmacy College. After we married, she had joined me in Princeton, where we sent our children to the preschool opened by Mrs. Oppenheimer, wife of J. Robert Oppenheimer, the head of the laboratory for the Manhattan Project. (After the war, Dr. Robert Oppenheimer became Director of the Institute for Advanced Study in Princeton until 1966.) To get to the preschool, we passed by the front of Einstein's house on Mercer Street every day. Akemi even went once to the physicist Dr. Wheeler's house; it was on an occasion when Mrs. Wheeler invited many ladies from foreign countries to her house near the Institute, and our friend Mrs. Shimada took her there. Dr. Wheeler had worked closely on the Manhattan Project, and had been physicist Richard Feynman's thesis director at Princeton.

In Princeton, therefore, we remained close to the memories of the atomic bomb. Then Tsutomu, when he had left Caltech, where he had studied with Richard Feynman, who had also worked on the atomic bomb in Los Alamos, decided to work at Los Alamos National Laboratory himself, to our great amazement! All these later events held greater significance for Akemi in view of her atomic bomb experience. Neither of us would ever forget the destruction in Nagasaki.

It was on the morning of April 17, two days after the Boston Marathon bombing, that my wife and I went through the tight security at Logan Airport in Boston, flying through Dallas, Texas, and continuing on to Santa Fe, New Mexico. Santa Fe, as the capital of New Mexico, is also a popular place for sightseeing. It is a very old city with 400 years of history behind it. But its airport is surprisingly small; after we disembarked from our plane, we only had to go a few steps to the arrival lobby where we met with Dr. Pearson and my son.

We headed to Los Alamos in Tsutomu's car. It was already dark and the frost was quite pretty—snowflakes caught on the tree twigs, looking like attractive flowers, almost as delicate as Japanese cherry blossoms. The road sloped upwards. I saw a steep valley on our right side, resembling the Grand Canyon. Los Alamos was chosen for secrecy by its first director, Dr. Robert Oppenheimer, more than seventy years ago (in 1942), and is understandably located in such a landscape.

Upon arriving at Los Alamos, we knocked at the door of the Hill Top Hotel at which we had made a reservation. A man came out and said that the hotel had closed half a month ago, in spite of the fact that we had a confirmation for our reservation. So Dr. Pearson quickly found another hotel for us.

The next morning, we looked out the window and saw the place by daylight for the first time. We found the scenery of Los Alamos rather strange. The town was situated at a high altitude of 2200 meters, and the surrounding mountains were brownish in hue, with white snow-covered mountains visible in the distance. We learned that there were two mountain fires in 2000 and 2011, which forced people to evacuate for two weeks. The hills surrounding Nagasaki were also brownish after the atomic bomb blast, but there the land turned green again before next summer. Probably the recovery of the land is slower at Los Alamos due to its high altitude and dryness.

Next morning, we were taken to the security office of the lab. They checked us even though they had already been investigating us during the last three months. After having our photographs taken, we received badges to use during our time in the lab. I saw many people waiting to get a badge; they were apparently not scientists. Cameras, telephones and electronics were all off limits there. Mr. John Markoff of the *New York Times,* who came to hear my lecture, had to be accompanied by an official at all times.

10.5 Fragments of History?

As we checked in and talked with people at Los Alamos, I wondered, from the perspective from the lab, *were the atomic blasts in Nagasaki and Hiroshima merely significant as a fragment of history?* First, the vice-director, Dr. McBranch, explained the history of the lab to us. In 1942, to make an atomic bomb in secrecy, Dr. Oppenheimer chose this place and purchased an existing summer school to make into the lab. The original facility was stuffed with about two hundred people. According to old pictures, the scientists lived in a wooden barracks, drying their laundry outside.

It must have required tremendous effort, and it was an amazing accomplishment, to gather people with the highest brainpower who could live in such a place to make an atomic bomb in two years. For any correspondence with the outside world, they had to use the post office in Santa Fe, a distance of two hours' drive by car, and all their correspondence was kept under a security watch. In Princeton, New Jersey, where the University and the Institute for Advanced Study exist, physicists were ordered to buy train tickets nearby at various stations, not at the Princeton station, to maintain secrecy; however, they all sent out their luggage from the Princeton station. We had read that the famous physicist Richard Feynman took his wife with tuberculosis to a sanatorium in Albuquerque and visited her every week. He wrote that he often had a punctured tire on the way.

Los Alamos is currently staffed with about 10,000 people. While the laboratory continues research on weapons, most scientists are working on bio-science, chemistry, drugs for tuberculosis and cancer, and theoretical science. The green fluorescent protein I discovered is also widely used there.

We had a lunch with the representatives of scientists of various fields, except for weapons research. I asked a question to them, "It seems that what you are doing all looks like peaceful, basic science. Why do you need such high security?" The answer was, "No, we don't need such high security." Then I wondered, *why didn't they separate basic science research from weapons research?* I suppose that the budget for basic science research is benefited by doing weapons research. It is unfortunate that basic science research does not get as readily funded, when the possibilities for the future applications and advantages gained from basic science research are endless.

My lecture began at 1:00 in the afternoon. Its title was "The discovery of green fluorescent protein GFP was unexpected." The contents were nearly the same as my Nobel Prize lecture, but this time, I emphasized that it was almost a miracle because the path I took included at least seven alternate branches, and GFP would not have been discovered if even one of those branches had led me a different way.

One of the alternate ways I took was that, in 1965, I moved to a position in the United States where I had to obtain a research grant, resigning a stable permanent position at Nagoya University in Japan, in order to do my research on the luminescence mechanism of aequorin. Another important reason for my pursuing this route was the pull I felt to research the green fluorescent protein discovered during the purification of aequorin.

I explained in my lecture that, while presently I am doing research on bioluminescence, I had not started out with that intention or expectation. (Perhaps, under other circumstances, I might have become an engineer.) After the war, I had no school to enter for two and a half years. Then, the Pharmacy School of Nagasaki Medical College destroyed by the atomic bomb opened its new campus near my home, using the old buildings of a military pilot training facility. Therefore, I entered and graduated from that school. However, I had not much interest in pharmacy, but since the basis of pharmacy is chemistry, I chose to study chemistry. That "accidental" background in chemistry and the suggestions given me by my mentors led me eventually, by chance or fate, into my focus on biochemistry, bioluminescence, aequorin, and GFP. It was an unexpected path.

After my lecture, I met with Dr. Geoff Waldo. He does research on split GFP, and holds many patents on its application. According to his words, the uses and applications of GFP have spread worldwide, like a "tsunami," after he started his research. This tsunami, however, had its long-ago tiny origins within the mushroom cloud over Nagasaki, which eventually led me into the study of chemistry. But the path to the future was never a clear road.

Then I met with Dr. Pearson's theoretical science group. We were invited for dinner that evening at the best restaurant in town, called "Dixie Girl," but the place was so noisy that I had trouble hearing the conversation. Next morning, for two hours starting at 10:00 a.m., I met with graduate and postdoctoral students. There were not many sharp questions. For example, I received questions such as: What happened to the jellyfish after their outside ring was cut off? How did I feel when I received a call from Stockholm about my Nobel Prize?

There was no question about the atomic bomb blast in Nagasaki. It seems that is merely a small fragment of history for younger people.

After finishing the official events, we had lunch, then visited the Bradbury Science Museum. The museum was built in 1963 by the second director of the laboratory, Dr. Bradbury, specifically for the atomic bomb and related matters. The bomb called "Fat Man" that destroyed Nagasaki is much larger than the "Little Boy" that destroyed Hiroshima. There, we were able to see display models of actual size. The size of Fat Man probably meant that it could be observed from 10 kilometers away.

On that unforgetttable day—August 9, 1945—I saw the first B–29 that dropped a parachute with an instrument, probably to collect the explosion data. When the second B–29 had reached a position above Nagasaki, I had gone back into the building, and sat down on my stool. At that moment I was assailed by an extremely strong flash of light. If I had continued to look at the sky, it probably would have damaged my eyes.

Here, at Los Alamos, I saw the same parachute with the instrument on display that I had seen seventy years ago against a clear blue sky. There was an explanation of the explosion mechanism of the atomic bomb given, as if there was no secret involved. It seemed as if it could be made almost anywhere if the materials were available; I was a little frightened at the idea.

In Nagasaki in 1945, it had been days before people learned what the bomb had been. Details were not publically known, and even the extent of the damage was hard for people to assess because of disruptions to transportation and communications. Afterward, the injuries to people and the dangers of radiation were poorly understood because there had been no atomic bombs before. No one knew the full consequences. More than half a century later, Japan was still collecting data on the long-term effects: Periodically, when we visited Japan, Akemi was supposed to see doctors who were continuing to trace the health conditions and history of people within a certain radius of the impact site.

How much do we know, and how much do we really understand? I wondered, as we visited this place where science had accomplished a terrible wonder. John Markoff wrote a story for the *New York Times* (May 12, 2013) about our visit. In his story, he noted Dr. Pearson's comment that the person who started explaining the instrumentation attached to the parachute was dumbfounded to hear that I had actually seen the drop of the instruments. Perhaps they are just old history preserved in a museum, for many people, even scientists, who live in a post-atomic world.

After seeing the atomic bomb displays at the Bradbury Science Museum, I felt that a cloud in my mind became a little bit lessened. My wife, who lost her elder brother and all her father's business to the atomic bomb in Nagasaki, was more emotional at being reminded of her experiences of seventy years ago. As a scientist, she was also amazed at

By the road to the Trinity test site (2013).

Left: with Akemi by Fat Man replica; right: parachute (Los Alamos, 2013).

the Manhattan Project scientists' ability to make the bombs in such a short time (in just two and a half years) in the deep mountains, starting with the construction of the road and the laboratory building.

Akemi was nine years old at the time when the atomic bomb blasted Nagasaki and she was only 2.5 kilometers (about a mile and a half) from the center of the blast; therefore she was exposed to strong radiation. She was not physically injured, but suffered later from various illnesses due to her damaged immune system. Her eldest brother, Akira, was a medical student at the Medical College of Nagasaki University and his body was found a half year later in the bushes on the mountain near the college. He was around the same age as me. Both his name and Akemi's contain Japanese *kanji* related to the concept of light, which is sometimes written with the *kanji* characters for the sun and the moon next to each other. (While often "Akira" is written with that same character, Akemi's brother Akira's name had a different character containing only "sun" in part of his kanji name.) It is ironic that the atomic bomb, which caused his death, created such a bright flash of light—a light whose destructive power darkened or altered the paths of so many people.

10.6 Akemi's Memories of the Atomic Bomb

On the day of the atomic blast, my wife was not even supposed to be in Nagasaki. During the last few years of the war she had been living with the Honda family, in her uncle's house in the countryside, as many Japanese children did in those days to avoid the bombing.

Her cousin Kiwako was accepted to one of the colleges in Nagasaki that summer and her classes were going to start in the middle of August. Kiwako and her mother decided to go to Nagasaki, with a hired taxi to carry a lot of luggage, and Akemi accompanied them, hoping to have a chance to see her mother, who was living in Michino, a suburb of Nagasaki. The taxi arrived around 10:00 am on August 9 at the Nagasaki railroad station. They brought Akemi to a family friend's house near the station to be taken care of there, while Akemi's father, her cousin and her cousin's mother left for errands to prepare for college. It was a little before 11:00 a.m.

Akemi was talking with the fourteen-year-old daughter of the house, Sachiko Yokoyama, in the entrance hall at 11:02 a.m. Then suddenly a bright lightning-like flash blinded her for thirty seconds or so. When her sight returned, broken glass and the window frames from which the panes had fallen were scattered all over the tatami mats around her. They realized this was an air raid, so Mrs. Yokoyama, Sachiko's mother, told her daughter to take Akemi to the shelter immediately; the usual arrangement would have been for Sachiko to take her grandmother instead of Akemi. The first shelter they arrived at soon became endangered by the approaching fire, so they had to escape to another one, then yet another one. Between the escapes they also had to hide from the B–29s that occasionally appeared in the sky. One of their hiding places was in a cemetery; removing its stone, they entered underground where pots of cremated bone were kept. They continued to run from place to place, looking for safety almost all through that afternoon.

We later found out that these B–29s were not present for another raid, but only to check the aftermath of the atomic bomb blast. However, no one knew that at the time, so they contributed to the fear and confusion in Nagasaki. Akemi remembered seeing the city burning with flames in many places and a big dark red dirty sun in the hot summer sky.

Akemi's cousin Kiwako's group was walking on the road at the time of the blast at 11:02 a.m., and many things like roof slates fell down from the surrounding buildings, but they were able to hide in a roadside ditch and were not harmed much, fortunately. Akemi's father Mr. Okubo took them to his brother's house at the outskirts of the city where the damage was less. After he deposited the two there, Mr. Okubo went back toward the train station to search for his daughter Akemi. On the way many houses were on fire. The friend's house was still there, and he found the old grandmother staying there alone. She told him that everybody had gone to the shelter at the top of the hill, but she had decided to stay home on account of her weak leg. He recommended her to escape the house immediately to seek safety from the fire, and then left hurriedly to continue searching for his daughter. He went to the shelters, but none had survived the burning. He went to the military headquarters where Sachiko's aunt was working and left a message that if his daughter should appear there, to please send her to his brother's house a mile away where the other two were already safely staying. Then he left to continue his search.

As he guessed would happen, Akemi and Sachiko arrived at her aunt's office in military headquarters in the late afternoon. They were kept safe in the shelter, finally; they had spent most of the day frantically going through the city. The two girls were given a glass of water and a rice ball when some other people who had escaped the devastation and the destroyed shelters came to that place. In later years, when she made rice balls for a snack, Akemi would tell Sachi this story: a woman gave her an *omusubi* (rice ball), flavored probably just with salt, and it was the most delicious meal in her life, after having no meal since the early morning during their desperate search for shelter.

As Akemi knew that her father was looking for her, running around the city, Sachiko and Akemi soon started to walk again, heading this time to her uncle's home. The road was a heap of rubble, and electricity poles were down, so that walking was not easy and travelling took time. By following the instructions given to them and Akemi's childhood memory, the two girls finally arrived at her uncle's house and saw everyone sheltering there. Later Akemi's father came back and was relieved to know that she was safe. Then he immediately left for his own home on the other side of the city, across the epicenter of the blast. To avoid fires he took the railroad tracks as much as possible. Actually many people were using the tracks to escape, instead of the road which was impassable with rubble. After he returned to his home after midnight, he found that Akemi's older brother Akira had not come home that night on August 9, the day of the blast. Thus, he went back to the epicenter the next morning to search for his eldest son, carrying a canteen of water. He already knew that the most damaged part of the city was near the university. During the search he encountered one of the medical students with a wound at the roadside who happened to be a son of the Yokoyama family, Sachiko's elder brother; but he was unable to find his own son.

After three days of searching in vain, he and his wife (Akemi's parents) visited the site of Akira's classroom just 500 meters from the epicenter and found there most of the students' bones lined up just as they had been attending a lecture in the classroom. No one would have survived there. Assuming that their son must be one of them, they picked up some ash and a button of the same design worn by their son and took that home.

Most of his classmates perished in their classroom almost instantly and thus could not identify each other. Later estimates suggest more than 70% of the medical students died that day, along with many of their teachers. This loss must have contributed to the shortage of medical aid following the dropping of the bomb. Besides, no one understood the full consequences of radiation exposure.

Akemi's parents still looked for their son's whereabouts, praying for good luck. They were fortunate: in February after the end of the war, they received a call from his high school that their son's body had been found upon the mountain near campus where his father had searched for him for three days. They learned that their son had taken a shortcut to the top of the mountain, trying to escape to safety to the rear of the mountain, where his uncle lived. This shortcut road was only used by the electric company to maintain high-voltage electric poles. That was the reason his father had not found him during his search.

By then, his body had turned almost to bones which were wearing shoes. A kind electrician had checked carefully and found that his name and school name were on the shoes. Getting a notice from the electrician, the high school teacher had remembered Akemi's brother and realized whose body it was, and called my wife's parents. Akemi's parents were among the lucky ones. His parents removed his wristwatch from the body and then cremated it at that place. His mother later donated this watch to the museum founded to memorialize the fall of the atomic bomb in Nagasaki. The watch's hands had stopped, their fixed position still recording the time of the bombing.

Much later, Akemi's mother and her experience with the atomic bomb and the loss of her son appeared in a short documentary film made to accompany an exhibition titled *Nagasaki Journey*, which opened in 1995. I had a chance to see the exhibition at Harvard University. However, my mother-in-law passed away a year before and could not see it, regretfully.

While her father was searching the mountain, Akemi stayed in the shelter for three days, and then her aunt Mrs. Honda decided to go back home by the train, which had started operating from Michino, two stations away from Nagasaki. To go to the station, she and her companions had to cross the city through the epicenter, but there was no other way. On the morning of August 12, with a canteen of water and thick cotton-filled hoods to protect against expected air raids, the three of them, guided by Mrs. Honda's brother, who happened to live near the uncle's shelter, left for Michino station. The rubble was fairly cleaned out and the major roads were passable by then. But the city was all black and flat, still smoldering in many places. Some of the smoke came from cremations and the smell was indescribable. On the way to the station, Akemi could still see some of the burned bodies lying like big chunks of charcoal, those figures showing their struggle to their final minutes. Those scenes are hard for her to recollect; she has not talked about that much, as with everyone who saw those scenes of hell.

After they passed through the hardest hit places and had neared the railroad station, they encountered a black car, which was the taxi they took three days ago. Kiwako's father had worried about his wife and daughter and had hired the same taxi to look for and pick up his family. The taxi had started in the early morning and the driver was still looking for them. When they saw him, they felt as though they had met an angel in hell.

Akemi's parents' house was near so that they decided to stop by the house to say hello to her mother before returning home. The house was located two miles away from the epicenter and did not burn down, but pieces of broken roof and glass filled its yard. Akemi just said hello to her mother in the entrance; her mother was very glad her daughter had survived the incident. Food was scarce because no stores had opened, and even before the bomb, people had experienced great difficulty finding enough supplies to feed themselves. Akemi's aunt offered to take Akemi's younger brother and sister with her to the countryside where the food was more plentiful and it was safer for children. In the car, three adults and three children were crowded in. The driver had to drive the car back through the epicenter to Nagasaki to deliver Kiwako's uncle who had kindly escorted his relatives to Michino.

As mentioned before, the main road was mostly cleared, but still many small pieces of debris were on it; therefore the car had flat tires several times. The driver had to repair them repeatedly as they passed through Nagasaki until they had travelled far away from the epicenter. On the way it was hot inside the car, but if they opened their windows the passengers had to breathe the unpleasant smell of cremation.

They arrived at their home around midnight and Akemi's uncle was relieved after four days of anxiety. Akemi too was relieved, but their tiredness persisted for the rest of that summer. Their physical condition must have been very poor. Their bodies had been damaged by radiation, but no one thought of such ideas at all at that time, since the long-term effects of atomic weapons were not yet understood. They were inside of a 1.5–2 mile radius from the epicenter at the blast time, then passed through near the epicenter twice, three days later. A little more than one month later, Akemi and her brother contracted diphtheria but were able to survive the disease. Akemi's aunt contracted tuberculosis soon and died from it the next year. Her cousin Kiwako did not get sick at that time, but later she had complaints in many parts of her body for the rest of her life and suffered a miscarriage of her only child. She died of liver cancer at age 70. Sachiko Yokoyama, who helped Akemi escape from the bomb, is still alive, and Akemi always feels thankful to her.

10.7 Science and War: Back to Nagasaki

After finding there, in the museum at Los Alamos, a picture of President Truman, I told Dr. Pearson that I hated him for dropping the second bomb in Nagasaki only three days after the first one was dropped in Hiroshima. I meant that the effect was not completely clear in three days. Could he not have waited longer to see if the second bomb was truly necessary? Dr. Pearson understood what I meant, but said that all the scientists in Los Alamos did not want to use the bombs in actual war, and the main purpose of the second bomb was to prevent the occupation of Japan by Russia, by eliciting the quick surrender of Japan.

Triggering the explosion of Little Boy, which used uranium, was relatively simple, and there was no need for a test in advance of its actual use. However, the test of Fat Man, which used plutonium, was carried out on July 16, 1945, at the Trinity test site in New Mexico. The explosion was much stronger than expected. A scientist there said, "We scientists are sons of bitches!"

I cannot criticize the use of the atomic bomb in Hiroshima strongly. If Japan had had it, Japan probably would have used it. I am sorry that Japan had not surrendered two weeks earlier; then the surrender would have saved a great many people. However, all were results of a war. I wish to have the world without any war. And I wish to have a world with no nuclear weapons.

Between the time that I received the invitation to Los Alamos and the time at which we were able to fit the visit into our schedule, we made another trip to Japan. Akemi thought that our granddaughter, who had been born in 2009, the year after the Nobel, was old enough to enjoy seeing Japan now. This time my visit included a stop at the Nagasaki Atomic Bomb Museum. Akemi chose not to go to the museum this time, so I offered to take Sachi there. One of the results of the Nobel award is that I am often recognized in Nagasaki. When I took Sachi and her family, including her husband and their three-year old daughter, to the Atomic Bomb Museum, I managed to remain anonymous on the streetcar ride. Perhaps it was because my granddaughter attracted all the attention from a group of school kids, who heard her chattering and wanted to try out their English on her. They held a simple conversation—"What is your name? How old are you?" and she tried out some Japanese. (Her favorite word was *cho-cho* 'butterfly,' and she liked to wave her arms like wings, causing passersby, especially old ladies, to comment on how *kawai* 'cute' she was.) I think the school children taught her a few more Japanese words, and told her she was cute. They all enjoyed the interaction, and were sorry to part when we reached our stop.

At the museum, a person at the entry recognized me and gave us special access, and later the director of the museum offered us a short tour and invited us to his office. The museum features a lot of water and translucent effects in honor and memory of the terrible thirst felt by the burned victims of the blast. There were also streamers of origami cranes. Sachi had asked to see again the watch that had belonged to her mother's brother who had died in the atomic bomb blast, and the director kindly arranged for it to be brought out from storage; that watch had made a strong impression on her when she had first visited the museum, probably as a teenager. She remembered seeing it in the glass case displayed next to the notebooks in which he had drawn sketches of B–29s and other aircraft. In the current rendition of the museum, we could pull up a picture of him in the database of bomb victims. He looked very young to me.

After our visit, my brother-in-law Akihiro came to drive us back in his car. We stopped at the Peace Park to see the statue. It is a massive seated bronze figure, nearly ten meters high, with one arm pointing up and the other extending outwards, and dedicated to the hope that no such devastation occurs again. There a group of school children who were on a field trip with their teacher recognized me, and asked for autographs and pictures; it was a busy several minutes. My granddaughter ran off to see the end of the park. The day was humid, as is typical in Nagasaki, but beautiful, with the sounds of children and the smell of late spring. Again, a nearby fountain served as a memorial to the thirst suffered by victims in the aftermath of the blast. Close by, a shrine contained strand upon strand of ribbons strung with a thousand paper cranes.

Later, Akihiro fed us all with large ice-cream bars; these were one of my granddaughter's fondest memories of Japan! As we sat, I thought about the past and the contrasts that lived in my remembrance.

Two other trips hold a particular place in my memory. On the small table by my granddaughter's bed, a small pink furry jellyfish rests upon its plush tentacles. It has a cute expression, and big eyes, and its name is Lily. It was a gift to me from the Aquarium in Sasebo, Japan. Later I was invited to the opening of the jellyfish exhibit in another aquarium in Tsuruoka in northern Japan in 2014. It seems my research has made jellyfish more fashionable in Japan.

The next year, I returned again to Japan. One of the events I was invited to was the Pugwash Conference. In January of 2015 I had received a letter from the organizer of the 61st Pugwash Conference on Science and World Affairs inviting me to participate in it. It was interesting to me because it would be held at Nagasaki in November of that year, upon the 70th anniversary of the atomic bomb. It was titled "Nagasaki's Voice: Remember Your Humanity," and called for the abandonment of nuclear weapons. The Pugwash Conferences' efforts had been recognized by the 1995 Nobel Peace Prize, and many Nobel laureates have supported their call for a world without the horrors of nuclear destruction.

However, my old age prevented me from participating completely through the conference; I only accepted the Dorothy Hodgkin Lecture. I chose to focus in my lecture on my bitter experience with the atomic bomb in my youth alongside my Los Alamos visit in 2013 to learn about the making of the bombs in the Manhattan Project 70 years ago, and to see the present laboratory. Physics laureate Toshihide Masukawa gave a few comments after the talk. He emphasized that he was struck by how my family had helped with my research, especially in collecting specimens, whereas his field of particle physics was a much more solitary endeavor.

Science, solitary or otherwise, continually connects back to the world in unforeseen paths. After this event I felt my life as a chemist had converged to an appropriate end: I started my life of being a chemist because of the atomic bomb at Nagasaki 70 years ago and now in this Pugwash Conference I realized I had had long enough of a life as a chemist.

CHAPTER 11

Epilogue

I would think that my life so far has been generally lucky; there seems to be more good luck than bad luck, with the Nobel Prize as the greatest of good luck. After receiving the Nobel Prize, I have been overwhelmed by the numerous requests from universities, scientific societies and media which have made it impossible for me to do my own research. I had no time at all to read scientific papers or to do experiments. During this time, the pile of research papers I should read has just continued to enlarge; under such conditions I cannot be an active scientist.

The effect of receiving a Nobel Prize was huge. Our lives changed. The focus of my work shifted. However, I am not sure yet whether it brought me good luck or happiness overall. In the past, I experienced a great joy two times, both times arising from my scientific pursuits. The first occasion was when I succeeded in the crystallization of *Cypridina* luciferin in 1956, and the second was when I found how to extract aequorin from the jellyfish *Aequorea*. It is always joyful when an experiment or undertaking has worked well, but those two times were especially so: I thought I was now in the group of real scientists, and had the confidence to go ahead as a scientist. I was really happy. The Nobel Prize was good luck that gave me a great honor, but it was not good luck that could give me such joy and happiness.

However, the most important matter for a scientist is the results of one's research. I am really satisfied that my research results have contributed sufficiently to the development of science so as to be awarded a Nobel Prize.

11.1 Was the Discovery of GFP Good Luck?

The most important item among my research results is probably the discovery of the photoprotein aequorin and GFP, both found in the jellyfish *Aequorea*. GFP and the various fluorescent proteins developed from it are now being widely used in biology and other fields as a marker protein; they are now indispensable in research.

The discovery of aequorin and GFP was made possible by various factors including the guidance by my three mentors, some accidental happenings, assistance by many people, the developments in the sciences, especially of genetics, and—finally—the efforts undertaken by myself. GFP and aequorin would be unknown today if those factors were different even slightly. The discovery of GFP was a rare piece of luck that nature granted to humans. The luck occurred because I picked it up when it came before me.

I would like to take this opportunity to emphasize the importance of learning from nature. Humans have learned various scientifically important matters, including genetics, by investigation of the various occurrences and mechanisms in nature; namely, we have learned from nature. Considering that humans were created by nature, humans cannot surpass nature; therefore, probably it would seem impossible to completely understand nature. Despite centuries of study, we still cannot explain many of its aspects. Thus it would be best that we remember to study nature in order to obtain new scientific knowledge.

11.2 My Research Life and Communications with Other Scientists

I prefer to do research that searches for the true foundation of things, rather than trying to find applications and usage. It is often an unappreciated effort in the background. I am not sure whether I was born this way or if my preference was caused by my earlier life. However, the idea of a *samurai* spirit or manner that was taught by my grandmother when I was young probably influenced me at least to some extent.

On looking back upon my research life, it seems that the way I chose was not really decided by myself. I just took the way shown by mentors. It began in 1955 when I studied at the laboratory of Professor Yoshimasa Hirata in Nagoya University by the arrangement of Professor Shungo Yasunaga of Nagasaki University where I was an assistant.

Professor Hirata gave me as a research subject the crystallization of *Cypridina* luciferin. Then, as the result of successful crystallization, I began to study bioluminescence. I was invited to the laboratory of Professor Frank H. Johnson, at Princeton University, to do research on the jellyfish *Aequorea* and other bioluminescent organisms. Therefore, I had taken the way shown by my three mentors, but the way I took was not an easy one as I explained in these present writings.

In studying bioluminescence, which has not been well studied, I had no established strategies for investigating a new bioluminescence system. Consequently, when studying a new bioluminescence system, one must develop or create the method for investigating it, which is difficult to achieve without considerable setbacks and obstacles. Generally speaking, modern scientists have been trained to use and develop knowledge already known and are often poor in creativity, and thus their training is not suitable for a wall-breaking job. In my case, therefore, I could not expect help from other people except on the practical peripheries, like collection; on the core research, I had to work by myself. Thus, I had to plan and work alone. It is lonely research. That is not my wish: it is a consequence.

Without the need for discussions with other people, I thought and carried out experiments alone, often without talking. As a result, I gradually found it more difficult to speak English, as well as Japanese. It was a necessary thing to achieve an objective. However, writing English was a different matter. Because the results of research are usually published in English, I must be able to write correct and accurate English. In connection with this matter, the advice Dr. Johnson gave me was extremely valuable: Always check a dictionary if you have any questions.

I don't remember having any competition with other researchers. I had struggled with Marlene DeLuca in the early 1970s on the reaction mechanism of the firefly luminescence, but it was to correct an error in her paper, not a competition to see who succeeded first in research. In the chemistry of bioluminescence, there are still a great many subjects available for rather a small number of researchers. Another reason for my lack of competition with other scientists is because I dislike to study popular subjects.

11.3 The People I Respect

There are many great people in history. But, here, I limit my thoughts to those whom I personally know. There are two categories of respect: in personality and in science.

First, I honor and respect my three mentors. All of them took care of their students, probably in the effort of developing their students to carry on worthwhile research. In the case of Professor Yasunaga, he took me from Nagasaki to Nagoya to introduce me to Professor F. Egami, probably at his own expense. Usually writing a letter of recommendation or introduction would be sufficient for the purpose, but he probably wanted to make sure of the results. Without that trip, I would not have met Professor Hirata in that twist of fate that I have described earlier.

When I decided to go to Princeton, Professor Hirata undertook, on his own initiative, the necessary actions to give me a PhD degree, which I was not expecting. Probably he thought that I should have that degree to succeed in the United States, and he filled out all the needed documentation for me by himself. Therefore I had never seen that documentation until the time when the mass media found it at Nagoya University after I received my Nobel Prize. Without his support, I may not have had the groundwork and credentials to win funding for my research in the United States.

Dr. Frank H. Johnson was a typical Princeton gentleman. He had always supported my research with his best efforts for fifteen years until his retirement. He supported my work even when he had disagreed with my approaches. He considered my future while planning to retire. He and his wife befriended me and Akemi, and generously helped us embark on our life in the United States.

I think my three mentors were great persons; I doubt if I can behave like them.

In addition, there are many people whom I respect in science; however, I especially respect three persons: Drs. E. Newton Harvey, Shinya Inoue, and Yoshito Kishi. Dr. E. Newton Harvey (1887–1959) created the field of bioluminescence from almost nothing. His book (*Bioluminescence*, 1952) is considered to be the Bible of bioluminescence. Dr. Shinya Inoue, who is at the Marine Biological Laboratory, like me, is the inventor of the Shinya Scope, and made a great contribution to the development of the polarization microscope and imaging methods. Dr. Yoshito Kishi is well known for the synthesis of complicated natural products. I have done several collaborative works with him that are rare undertakings for me, and have consulted with him occasionally. I respect him as a person who makes the most chemically accurate and reliable judgments.

It has also been my fortune to meet and sometimes work with or beside many other wonderful people and great scientists.

11.4 A Word of Thanks

Alongside my mentors and colleagues in science, I am thankful to many people on this long path—many more people than I can name here. Finally I am thankful to my wife and children who helped my research. Also I am extremely grateful for my family and close relatives in Japan who made me continue my research by helping me any way they could. Some of my story has been about them.

Selected Bibliography

This bibliography contains only those works that are most relevant to my biography. For a fuller list of materials relevant to bioluminescence, consult O. Shimomura (2012), cited below:

Head, J. F., Inouye, S., Teranishi, K., and Shimomura, O. (2000) The crystal structure of the photoprotein aequorin at 2.3Å resolution. *Nature* 405: 372–376.

Shimomura, O. (1979) Structure of the chromophore of *Aequorea* green fluorescent protein. *FEBS Lett.* 104: 220–222.

Shimomura, O. (2005) The discovery of aequorin and green fluorescent protein. *J. Microsc.* 217: 3–15.

Shimomura, O. (2006) *Bioluminescence: Chemical Principles and Methods.* World Scientific, Singapore.

Shimomura, O. (2009a) Discovery of green fluorescent protein, GFP. In: Grandin, Karl (ed.), *Les Prix Nobel. The Nobel Prizes 2008.* Nobel Foundation, Stockholm.

Shimomura, O. (2009b) Osamu Shimomura—Biographical. In: Grandin, Karl (ed.), *Les Prix Nobel. The Nobel Prizes 2008.* Nobel Foundation, Stockholm.

Shimomura, O. (2010) *Learning From Jellyfish [Kurage ni manabu].* Bunkensha, Nagasaki (in Japanese).

Shimomura, O. (2012) *Bioluminescence: Chemical Principles and Methods* (revised edition). World Scientific, Singapore.

Shimomura, O., Goto, T., and Hirata, Y. (1957) Crystalline *Cypridina* luciferin. *Bull. Chem. Soc. Japan* 30: 929–933.

Shimomura, O., Johnson, F. H., and Saiga, Y. (1962) Extraction, purification and properties of aequorin, a bioluminescent protein from the luminous hydromedusan, *Aequorea. J. Cell. Comp. Physiol.* 59: 223–229.

Shimomura, O., and Johnson, F. H. (1966) Partial purification and properties of the Chaetopterus luminescence system. In: Johnson, F. H., and Haneda, Y. (eds.), *Bioluminescence in Progress*. Princeton University Press, Princeton, NJ, pp. 495–521.

Shimomura, O., and Johnson, F. H. (1972) Structure of the light-emitting moiety of aequorin. *Biochemistry* 11: 1602–1608.

Shimomura, O., and Johnson, F. H. (1975) Regeneration of the photoprotein aequorin. *Nature* 256: 236–238.

Zimmer, Marc. (2005) *Glowing Genes: A Revolution in Biotechnology*. Prometheus Books, Amherst, NY.

Index

acrasin, 124, 125
Aequorea, 62, 70–76, 98–100, 108, 109, 121, 140, 142, 148–151, 159, 172, 203–205
aequorin, 41, 75, 76, 85, 87, 95–98, 101–103, 105–109, 113, 127, 128, 132, 136–138, 140, 145, 147, 148, 150, 151, 153, 189, 190, 203, 204
Arachnocampa, 83, 84
Asahi Prize, 152–154
atomic bomb, 27, 29–31, 35, 39, 55, 59, 61, 83, 94, 133, 185–202

Bergen, 121, 141, 143
Bermuda Marine Laboratory, 77
Bonner, Dr. John, 124, 125, 131
Boston University, 163
Brehm, Dr. Paul, 128
brittle star. *See* Ophiopsila, 128
Buck, Pearl S., 33, 34

calcium, 73–75, 85, 96, 97, 101, 102, 108, 109, 127, 128, 138, 147, 150
Chaetopterus, 76, 87, 88, 105, 108
Chalfie, Dr. Martin, 128, 138, 145, 147, 158, 170–172, 175
Chicago, 57, 58, 65, 74, 152
chromophore, 49, 97, 105, 113, 138, 139, 150
coelenterazine, 101–103, 107, 109, 128, 136, 137, 141
Cypridina, 41–44, 46, 48, 49, 52, 63, 71, 72, 82, 83, 85, 101, 102, 106, 113, 114, 128, 131, 132, 136, 151, 179, 203, 205

Dan, Dr. Katsuma, 115, 116
dinoflagellate, 114
dioxetane, 106
Dunlap, Dr. Jay, 114

Euphausia, 104

Fay, Dr. Rimmon, 76, 81, 87, 88
Flood, Dr. Par, 140
Friday Harbor, 62, 65–71, 73, 74, 76, 89, 94, 95, 98–100, 113, 116, 129, 147–149, 172, 173

Friday Harbor Laboratories, 66, 70, 71, 95, 113, 147, 148, 151
Fulbright, 51, 53, 57, 58, 75, 120, 140, 163, 181
fungi, 134–136

GFP, 74, 75, 87, 113, 128, 138–140, 145, 147, 150–152, 155, 158, 159, 167, 168, 178, 189, 190, 204
Gitelson, Dr. Josef, 182, 184
Goto, Dr. Toshio, 42, 45, 47, 49, 93, 106, 107, 128, 129, 131
green fluorescent protein. *See* GFP, viii, 75, 143, 168, 189

Hakon Mosby, 141
Harvey, Dr. E. Newton, 42, 49, 52, 63, 72, 84, 106, 131, 132, 207
Hastings, Dr. J. Woodland, 112, 114
Head, Dr. James, 137
Hikawa-maru, 53, 58
Hirata, Dr. Yoshimasa, 40–42, 44, 45, 48, 51, 78, 82, 87, 128, 132, 154, 179, 183, 205, 206
Hirata lab, 41, 42, 46, 51, 154, 179
Hiroshima, 27–30, 59, 61, 82, 185, 188, 190, 199

Iceland, 120–121, 140, 164, 165, 175
Inoue, Dr. Satoshi, 137
Inoue, Dr. Shinya, 99, 105, 112, 145, 147, 161, 207
Inoue, Dr. Shoji, 93, 101, 107
Isahaya, 11, 17–19, 21, 23–25, 28–35, 133

Johnson, Dr. Frank H., 49, 51–53, 59, 62, 63, 65, 71–74, 77, 83–86, 88, 92, 93, 95, 98–100, 104, 106, 109, 111, 112, 124, 129, 131, 132, 205–207

Kaminer, Dr. Benjamin, 112
Kanda, Dr. Sakyo, 63, 115
Kishi, Dr. Yoshito, 82, 83, 104, 127, 129, 207
Kristinebergs Zoological Station, 104

Latia, 83, 84, 103
Lienshankuan. *See* Renzankan, 7
Lindau, 180, 181
Los Alamos, 185–188, 191, 192, 198, 199, 201
luciferin, 41–49, 52, 53, 61, 63, 71–73, 77, 78, 82–85, 96, 101–108, 113, 114, 126, 128, 134–136, 140, 151, 184, 203, 205
Luminodesmus, 119, 125–127

Manchuria, 5, 7, 8, 12, 14, 15, 17, 30, 182
Marine Biological Laboratory, 60, 99, 105, 112, 115, 116–118, 123, 124, 145, 153, 159–163, 177
Masukawa, Dr. Toshihide, 158, 170, 171, 173, 201

Meganyctiphanes, 104
millipede. *See* Luminodesmus, 119, 125–127
Millport Marine Station, 104
Motyxia. *See* Luminodesmus, 125
Mycena citricolor, 135

Nabeshima, 8, 55, 56
Nagasaki, 8, 11, 24, 27–31, 33, 40, 57, 61, 83, 90, 94, 133, 143, 153, 167, 178, 179, 185–188, 190–194, 196–202, 206
Nagasaki Pharmacy College, 31, 35, 36, 38, 39, 48, 153, 154, 158, 186, 189, 205
Nagasaki University, 54, 56, 76, 78, 79, 133, 160, 178, 179
Nagoya, 40, 41, 43, 47–49, 59, 81, 83–86, 89, 93, 129, 154, 178, 179, 183
Nagoya University, 40, 58, 78, 79, 87, 154, 178, 184, 189, 205, 206
Nakamura, Dr. Hideshi, 104, 114
New Zealand, 83–85, 103
Nicolas, Dr. Marie-Therese, 114
Nobel Museum, 148, 166
Nobel Week, 157, 160, 163–166, 169, 170, 173, 175, 177
Norway, 121, 140, 142

Odontosyllis, 77, 78
Ohmura Naval Aircraft Factory, 24, 25, 30
Ophiopsila, 128
Order of Culture, 158, 160–162, 169, 170, 173
Osaka, 19–21, 153
Ozaki, Dr. Hironobu, 67, 68, 88, 95, 124

Panellus, 135
Parapriacanthus, 52
Parazoanthus, 76
Pearse Prize, 147
Periphylla, 128, 140–143
Pholas, 72, 109
photoprotein, 75, 108, 109, 128, 147, 185
Princeton, 33, 58–60, 63, 74, 76, 78, 81, 87–89, 91, 93–95, 103, 119, 129, 169, 186, 188
Princeton University, 33, 42–45, 48, 49, 51, 52, 55, 59, 74, 82–86, 88, 90–93, 100, 104, 106, 112, 115–117, 122, 124, 131, 132, 139, 145, 146, 148, 183, 205
Pugwash Conference, 201, 202

Renzankan, 7, 12
Royal Microscopical Society, 147
Russia, 182, 184, 199

Sasebo, 5, 8, 12, 13, 15, 16, 18–21, 179, 201
scale worm, 109, 114
Setoda, 82, 179
Shibusawa, Eiichi, 8–10
Shimoda Marine Laboratory, 49

Shimomura, Akemi, 32–34, 53–55, 57, 60, 61, 63–68, 83, 86, 88, 91–95, 99, 100, 111, 112, 116, 119, 120, 122–123, 125, 128, 129, 131, 133, 134, 141, 142, 147, 152, 155, 158–164, 168–172, 174, 186, 191–202, 207
Shimomura, Chikara, 5, 9, 11, 32–34
Shimomura, Tsuki, 6, 8, 12, 15, 16, 54, 133, 154
Shimomura, Tsutomu, 7, 61, 62, 83, 86, 87, 94, 95, 99, 111, 116, 118, 119, 122, 123, 158, 160, 165, 166, 168–170, 173–175, 183–187
Stockholm, 148, 158, 160, 164, 166–168, 170, 171, 173–175, 177, 190

Takeda, Matsue, 14, 17, 23, 27, 54, 56, 182
Teranishi, Dr. Katsunori, 108, 160
Tsien, Dr. Roger, 145, 147, 158, 165, 170–172, 175

Washington, DC, 75, 76, 164, 181
Watasenia, 84, 93, 101, 107, 108
Water Research Institute, 87
Water Science Institute, 78, 81, 86

Yasunaga, Dr. Shungo, 38–41, 132, 205, 206